RENEWABLE ENERGY: RESEARCH, DEVELOPMENT AND POLICIES



# U.S. RENEWABLE ENERGY AND ENERGY EFFICIENCY EXPORT INITIATIVE

# RENEWABLE ENERGY: RESEARCH, DEVELOPMENT AND POLICIES

Additional books in this series can be found on Nova's website under the Series tab.

Additional E-books in this series can be found on Nova's website under the E-book tab.

# ENERGY POLICIES, POLITICS AND PRICES

Additional books in this series can be found on Nova's website under the Series tab.

Additional E-books in this series can be found on Nova's website under the E-book tab.

RENEWABLE ENERGY: RESEARCH, DEVELOPMENT AND POLICIES

# U.S. RENEWABLE ENERGY AND ENERGY EFFICIENCY EXPORT INITIATIVE

KELVIN NORRIS
AND
CAMERON STEELE
EDITORS

Nova Science Publishers, Inc.
*New York*

For permission to use material from this book please contact us:
Telephone 631-231-7269; Fax 631-231-8175
Web Site: http://www.novapublishers.com



**Library of Congress Cataloging-in-Publication Data**

U.S. renewable energy and energy efficiency export initiative / editors, Kelvin Norris and Cameron Steele.
p. cm.
Includes index.
ISBN 978-1-61942-354-1 (soft cover)
1. Energy policy--United States. 2. Renewable energy sources--United States. 3. Energy consumption--United States. 4. Exports--United States. I. Norris, Kelvin. II. Steele, Cameron. III. Title: US renewable energy and energy efficiency export initiative.
HD9502.U52U184 2011
333.79'40973--dc23

2011046270

*Published by Nova Science Publishers, Inc. † New York*

# CONTENTS

# PREFACE

The Obama Administration believes strongly that the United States must lead the clean energy economy of the 21st century. When more of our electricity comes from clean energy, our economy will be more stable, our country more secure, our people healthier and our environment more sustainable. In addition, building a robust clean energy industry will help create the high-paying, high-technology jobs our country's future requires. This book examines the U.S. Renewable Energy and Energy Efficiency Export Initiative which the Obama Administration has launched and which is the most concerted effort to date to develop America's capacity for manufacturing clean energy technologies.

Chapter 1- The Obama Administration believes strongly that the United States must lead the clean energy economy of the 21$^{st}$ century. When more of our electricity comes from clean energy, our economy will be more stable, our country more secure, our people healthier, and our environment more sustainable. In addition, building a robust clean energy industry will help create the high-paying, high-technology jobs our country's future requires.

For this reason, the Obama Administration launched the most concerted effort to date to develop America's capacity for manufacturing clean energy technologies. Over $90 billion was invested through the Recovery Act to help double the country's installed capacity of renewable energy technologies and dramatically increase the country's clean energy manufacturing capacity. But as U.S. companies improve their technology and expertise, it is critical that the U.S. Government provide the export promotion our companies need to compete in growing markets overseas.

Chapter 2- The Trade Promotion Coordinating Committee Working Group on Renewable Energy and Energy Efficiency (RE&EE) is proud to announce that one year into the implementation of the Renewable Energy and Energy Efficiency Export Initiative (RE4I), every commitment has been achieved or is in the process of completion. Spurred by President Obama's challenge to position the United States as the leading exporter of clean energy technologies, the RE4I streamlined the process of obtaining financing for renewable energy exporters, dedicated additional resources to financing exports in the sector, and improved the delivery of existing export promotion services for RE&EE companies – all in an era of scarce resources and budget uncertainty. In addition, the RE4I created a process for improving market access for RE&EE exporters and better targeting future trade promotion activities.

Chapter 3- The United States and its trading partners have put a high priority on domestic and international efforts to reduce emissions of greenhouse gases, mitigate climate change, and protect the environment. In the face of these efforts, those countries that develop and foster competitive environmental goods industries will be in the best position to reap the economic rewards of increased international demand for goods that mitigate climate-related and other environmental challenges.

In December 2009, my office produced a special report entitled Major Opportunities and Challenges to U.S. Exports of Environmental Goods. The report showed that (1) global trade of environmental goods more than doubled in recent years; (2) although U.S. exports of environmental goods increased, the U.S. trade balance of these goods deteriorated substantially; (3) trade barriers in the fastest growing markets abroad are a key constraint to increasing U.S. exports; and (4) U.S. exporters are losing overseas market share to foreign producers of environmental goods, especially to China.

In: U.S. Renewable Energy and Energy ... ISBN: 978-1-61942-354-1
Editors: K. Norris and C. Steele 

*Chapter 1*

# RENEWABLE ENERGY AND ENERGY EFFICIENCY EXPORT INITIATIVE*

## *The Trade Promotion Coordinating Committee Working Group on Renewable Energy and Energy Efficiency*

## FOREWORD

The Obama Administration believes strongly that the United States must lead the clean energy economy of the 21st century. When more of our electricity comes from clean energy, our economy will be more stable, our country more secure, our people healthier, and our environment more sustainable. In addition, building a robust clean energy industry will help create the high-paying, high-technology jobs our country's future requires.

For this reason, the Obama Administration launched the most concerted effort to date to develop America's capacity for manufacturing clean energy technologies. Over $90 billion was invested through the Recovery Act to help double the country's installed capacity of renewable energy technologies and dramatically increase the country's clean energy manufacturing capacity. But as U.S. companies improve their technology and expertise, it is critical that the U.S. Government provide the export promotion our companies need to compete in growing markets overseas.

---

* This is an edited, reformatted and augmented version of The Trade Promotion Coordinating Committee Working Group on Renewable Energy and Energy Efficiency publication.

> "The nation that leads the world in creating new energy sources will be the nation that leads the 21st-century global economy."
>
> President Obama
> April 22, 2009

To that end, in January, President Obama announced the National Export Initiative (NEI), seeking to double exports in five years to support several million new jobs. At the Department of Commerce, we have redoubled efforts to eliminate trade barriers and enhanced trade promotion activities; and we have strengthened our commitment to promoting U.S. competitiveness in high-growth sectors like clean energy.

The following *Renewable Energy and Energy Efficiency Export Initiative* demonstrates this commitment. The Initiative is the country's first-ever attempt to coordinate U.S. Government programs in support of renewable energy and energy efficiency exports. Through the implementation of 23 new commitments from 8 U.S. Government agencies, the Initiative is designed to facilitate a demonstrable increase of renewable energy and energy efficiency exports over the next five years – helping to meet the goals of the NEI and President Obama's challenge to become the leading exporter of clean energy technologies. These actions will be implemented consistent with federal law and international agreements.

Developed through the Trade Promotion Coordinating Committee, the Initiative lays the foundation necessary to help U.S. renewable energy and energy efficiency companies take better advantage of current market opportunities, and orients federal government programs to better support U.S. global competitiveness in these sectors over the long run. The Initiative will support RE&EE exporters by offering new financing products, enhancing market access, increasing trade promotion and improving the delivery of export promotion services to current and future RE&EE companies.

In a challenging economic environment, all of the Initiative's 23 commitments for new programs, actions, and deliverables will be undertaken within existing budgets and existing authorities at no additional cost to the American taxpayer. I am confident that this Initiative will provide the coordinated and collaborative support our renewable energy and energy efficiency companies need to enhance their success in the global marketplace.

Gary Locke, U.S. Secretary of Commerce,
Chair of the Trade Promotion Coordinating Committee

## Abbreviations and Acronyms

| | |
|---|---|
| c-Si | crystalline silicon |
| CSP | concentrated solar power |
| EPA | Environmental Protection Agency |
| ETA | Employment and Training Administration |
| Ex–Im Bank | Export–Import Bank of the United States |
| FTA | free trade agreement |
| IGA | International Geothermal Association |
| IPR | intellectual property rights |
| ITA | International Trade Administration |
| HS | Harmonized Commodity Description and Coding System |
| HTS | Harmonized Tariff Schedule |
| kVA | kilo volt amperes |
| Kw | kilowatts |
| MW | megawatts |
| NEI | National Export Initiative |
| OECD | Organization for Economic Cooperation and Development |
| OPIC | Overseas Private Investment Corporation |
| RE&EE | renewable energy and energy efficiency |
| SBA | Small Business Administration |
| SWH | solar water heating |
| TPCC | Trade Promotion Coordinating Committee |
| USAID | U.S. Agency for International Development |
| USCTC | U.S. Climate Technology Cooperation |
| USTDA | U.S. Trade and Development Agency |
| USTR | Office of the U.S. Trade Representative |
| WTO | World Trade Organization |

## Introduction

In the 2010 State of the Union address, President Barack Obama announced the National Export Initiative (NEI), which is aimed at doubling exports within five years.[1] The ongoing implementation of the NEI calls on the Federal Government to facilitate exports in high-growth sectors, such as environmental goods and services, renewable energy, health care, and biotechnology, as a way to promote industries that can create high-paying jobs for the American people (see Box 1).

To help meet the goals of the NEI, the Trade Promotion Coordinating Committee (TPCC) established the Working Group on Renewable Energy and Energy Efficiency (RE&EE) and created the Renewable Energy and Energy Efficiency Export Initiative (Initiative). The Initiative is divided into two parts: (a) an assessment of the current competitiveness of U.S. renewable energy and energy efficiency goods and services and (b) an action plan of new commitments that facilitate private-sector efforts to significantly increase U.S. RE&EE exports within five years. Four annexes provide the following:

- Annex 1 defines RE&EE industries,
- Annex 2 provides a technology-specific analysis of RE&EE competitiveness and benchmarks,
- Annex 3 describes the underlying methodology of the Initiative's trade figures, and
- Annex 4 gives additional information on U.S. Government agencies, programs, and services.

## Part I: Competitiveness Assessment and Benchmark

The Initiative begins with an assessment of the competitive position of the U.S. RE&EE industries. Understanding the factors that contribute to U.S. RE&EE export competitiveness is critical to developing policies and programs to meet the needs of U.S. RE&EE companies. Some challenges identified in the competitiveness assessment, such as currency fluctuations and the presence of a domestic manufacturing base, are beyond the mandate of the TPCC Working Group on RE&EE. The commitments outlined in Part II address those barriers that are within the purview of TPCC agencies and can be accomplished within current funding levels. The Initiative's implementation is a first step toward significantly increasing U.S. RE&EE exports.

The Initiative also sets benchmarks of RE&EE exports, against which progress can be measured. In 2009, the United States exported roughly $2 billion of manufactured renewable energy goods (see Figure 1).[2]

Significant limitations exist in both the availability and the quality of data needed to accurately measure U.S. trade in RE&EE goods. The TPCC agencies, coordinating closely with the International Trade Commission, are developing reliable methodologies for acquiring and measuring export and import data. Yet accurate information on U.S. RE&EE exports will take years to develop. Therefore, policy-makers and stakeholders should consider this lack of substantial data before drawing conclusions based on the competitiveness assessment. As better data on RE&EE exports are developed, the Working Group will revise the 2009 benchmark.

Another critical challenge is the lack of data to track the exports of services related to the RE&EE industry. Services represent more than 70 percent of U.S. gross domestic product. Similarly, it is likely that services exports in the RE&EE industry are significantly larger than the exports of products, as tracked by this report. It would be a mistake to overlook opportunities to strengthen service exports such as architectural design of green buildings, energy audits, and licensing of U.S. wind turbine designs.

**Box 1. Who's Responsible for Promoting U.S. Exports?**

Companies in many European and Asian countries enjoy the support of a dedicated government agency that oversees all government resources promoting their exports. In the United States, at least 20 separate Federal agencies or departments, each with its own mandate, execute programs that support U.S. exporters. The Trade Promotion Coordinating Committee (TPCC) coordinates the development of U.S. Government trade promotion policies and programs.

Since it was formed in January 2010, the TPCC Working Group on Renewable Energy and Energy Efficiency has focused specifically on increasing the global competitiveness of the U.S. RE&EE sector. Twelve departments and agencies collaborated extensively on the Initiative, which aims to facilitate private sector efforts to significantly increase RE&EE exports over the next five years to help meet the goals of President Barack Obama's National Export Initiative.

## Part II: The Action Plan

The TPCC Working Group on RE&EE combined the results of the competitiveness assessment with the collective input of more than 50 companies, trade associations, and non-profit organizations gathered through a Federal Register notice and several consultations with industry stakeholders. The result was Part II of the Initiative, which is an action plan of agency commitments.[3] Eight U.S. Government agencies committed to address three common hurdles facing U.S. RE&EE companies seeking to export: a shortage of appropriate financing, limited market access, and a lack of contacts with qualified foreign partners and customers. These actions were agreed upon based on close consultations with twelve agencies participating in the working group. See Annex 3 for information on participating agencies. Part II also lays out a series of actions to strengthen existing U.S. Government programs. To address those obstacles, the Initiative lists four priorities: (a) tailor financing to RE&EE companies, (b) improve market access, (c) enhance information and trade promotion efforts to link buyers and sellers, and (d) strengthen U.S. government services.

*Tailor financing to RE&EE companies.* Financing allows U.S. companies to fulfill export orders and project developers to build power projects and to install energy-saving equipment. RE&EE companies face a wide range of obstacles to get financing, including several factors that are unique to the industry. To address this situation, several U.S. Government agencies will increase their funding for RE&EE projects, streamline the process for reviewing RE&EE funding applications, and enhance the scope and effectiveness of the products and services already provided to the RE&EE industry.

*Improve market access.* U.S. companies with exportable technologies and a desire to sell their products abroad need a concerted effort from the U.S. Government to open international markets to U.S. products and services.

Across all sectors of the economy, the government addresses trade barriers by (a) negotiating foreign import tariff reductions for goods and services, (b) negotiating multilaterally with other countries to clarify international rules regarding potential non-tariff barriers (NTBs) affecting goods and services, and (c) enforcing the outcomes of trade agreements by consulting with foreign governments and using dispute settlement procedures under the World Trade Organization (WTO) and other trade agreements as needed. The Initiative also outlines several actions aimed at helping countries with abundant RE resources that currently lack market mechanisms to create demand for RE goods and services and to develop the enabling policy framework necessary for trade in RE goods and services.

*Enhance information and trade promotion efforts to link buyers and sellers.* Exports often result from strong partnerships between U.S. companies and foreign buyers.

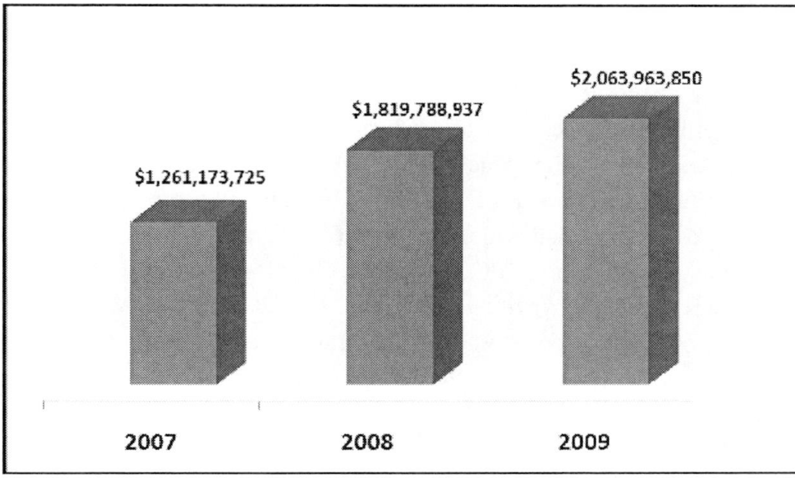


Source: International Trade Commission and U.S. Department of Commerce.

Figure 1. Renewable Energy Exports, 2007-2009.

One of the most important actions is to continue to strengthen, expand, and better coordinate the government's existing trade promotion efforts. However, a lack of information indicating which markets are most appropriate for U.S. exports hampers public and private efforts to effectively link buyers and sellers in the global RE&EE marketplace. Identifying markets that are most likely to result in U.S. exports will help maximize the effectiveness of U.S. Government services and will enable U.S. RE&EE companies to make smart decisions on market entry and expansion strategies. The Initiative will overlay a number of actions to enhance businesses' understanding of the RE&EE sectors and of potential foreign markets. The Initiative will give U.S. companies more opportunities to develop new connections with foreign partners and will identify markets that have strong RE&EE export potential.

*Strengthen U.S. government services.* Several existing U.S. Government programs are designed to increase trade and investment in emerging growth sectors, such as RE&EE. The Initiative outlines actions to strengthen and improve these federal resources and coordinate programs across agencies that would benefit both new and existing RE&EE exporters (see Box 3).

**Box 3. A Focus on Renewable Electricity Generation and Efficiency**

The RE&EE Export Initiative focuses on facilitating exports of goods and services related to renewable energy electricity generation and energy efficiency. Energy storage is considered in a limited way as it relates to electricity storage. All goods and services that relate to the transport sector, such as biofuels and biofuel feedstock, are not included.

The Initiative's scope was determined by the need to focus on technologies with similar issues and policy priorities. TPCC agencies are also undertaking a variety of initiatives to support clean or renewable technologies not included within the scope of this Initiative.

In addition to the TPCC RE&EE Working Group, the TPCC Civil Nuclear Trade Working Group is formulating its own strategic priorities to boost exports in this important clean energy sector. The RE&EE Working Group and the Civil Nuclear Trade Working Group collaborate closely with interagency partners to fulfill the President's goals in support of all U.S. clean energy industries, including the transportation sector.

# Part I: Competitiveness Assessment

U.S. companies benefit from increased exports in numerous ways. U.S. exporters profit from access to growing foreign markets, which allows them to achieve greater economies of scale and to lower their costs. During domestic economic downturns, exposure to foreign markets can help companies weather a recession and avoid layoffs. In addition, salaries in export-oriented industries are traditionally higher than those in the domestic market.

However, many U.S. RE&EE companies do not export, and those that do typically export to only one or two markets. Low export performance stems from numerous factors, including inadequate or non-existent market research, lack of manufacturing capacity, unfamiliarity with export logistics, risk aversion to foreign markets, lack of links to foreign partners or buyers, currency fluctuations, and an inability to finance sales abroad. Incomplete RE&EE regulatory and policy frameworks in foreign markets amplify those challenges.

Despite a global recession and uncertainty regarding the global response to climate change, $162 billion of private-sector capital was invested in

RE&EE technologies globally in 2009.[4] As the economy improves, global investment in RE&EE is likely to increase further.

Globally, nations are using RE&EE as a catalyst to promote economic growth and to create high-tech, high-wage jobs for their citizens. As of 2009, stimulus bills worldwide invested an additional $183 billion in RE&EE technologies.[5] The growth of foreign markets creates new export opportunities, even as foreign firms test the competitiveness of U.S. companies.

## The Importance of Domestic Policies in Creating Export Capacity

Predictable and favorable public policies are essential preconditions for the development of a robust RE&EE manufacturing base. Firms from countries that have provided long-term incentives and have removed barriers to commercializing and installing RE&EE technologies are challenging U.S. companies, despite the historic investment of the United States in RE&EE innovation.

In contrast, U.S. federal incentives, while considerable, have not been sufficiently long term to encourage investment in manufacturing capacity. Periodic uncertainty about program expiration—and some periods when the incentives such as the production tax credit and the investment tax credit actually expired— discouraged investment in manufacturing capacity to serve the domestic market. Although federal incentives led to significant deployment of renewable energy capacity in the United States, frequent expirations or struggles to renew the credits did not provide the stability that manufacturers needed to make confident investments.

State policies are invigorating the market and should have long-term effects. Energy policies and financing incentives in different states have become major drivers of RE&EE. Many states have implemented renewable portfolio standards (mandates that designate a percentage of electricity generation that must come from renewable energy sources) or public benefit funds to support energy efficiency initiatives. As of June 2010, 29 states plus the District of Columbia had some type of renewable portfolio standards, and 18 states had public benefit funds.[6] A relatively strong voluntary renewable energy market also exists, which allow corporations or consumers to pay a voluntary premium to purchase renewable energy (usually through renewable energy certificates) or to install on-site renewable energy production.

From 2008 to 2010, state policies supporting RE&EE, plus the investment tax credit and production tax credit for renewable energy, led to unprecedented additions of renewable energy capacity and significant investments in energy efficiency. The sustained boom resulted in an expansion of manufacturing capacity, illustrated by the growth of manufacturing facilities associated with wind energy, from 13 facilities coming on-line in 2004 to 37 starting operations in 2009.[7]

## A Global Race for RE&EE Manufacturing Capacity

China, Europe, and other countries and regions provide a robust set of national mandates and incentives that have led to higher levels of investment in and a buildup of their RE&EE manufacturing capacity. Consequently, the ability of U.S. companies to export their products has been limited by the lack of a domestic manufacturing capacity, which other countries now enjoy.

Several countries are targeting incentives to potential manufacturers to create manufacturing jobs for their citizens. China offers tax holidays for companies that operate in special economic development zones and that qualify as “new technology enterprises.” Malaysia offers 100 percent tax holidays for solar manufactures for periods up to 15 years. Foreign incentives for localized manufacturing pose a competitive threat to U.S. exports if companies choose to manufacture and export outside of the United States.

## Access to Foreign Markets

Unfortunately, the desire for green job creation in foreign markets has encouraged misguided policies designed to give preference to domestic companies. Examples of such protectionist policies include high tariffs, local content requirements, and other preferential treatment for domestic companies.

Such policies may be inconsistent with obligations in the WTO or other trade agreements and will be addressed by the U.S. Government where they appear. They may not only limit RE&EE exports but also threaten to slow the development of efficient global supply chains. Local content requirements (policies that mandate the use of a specified amount of domestically produced goods or services) discourage U.S. exports by requiring that U.S. firms manufacture offshore to access a particular market.

In addition, barriers such as weak protection for intellectual property rights, non-transparent regulations and licensing regimes, and burdensome certification requirements are less obvious but equally restrictive to U.S. exporters. Protection of intellectual property rights (IPR) is also critical to providing U.S. companies the confidence needed to export innovative RE&EE technologies. IPR is especially important to new RE&EE companies that depend on innovation to stay in business. Inadequate IPR protection and enforcement mechanisms increase the risk for U.S. companies seeking to enter foreign markets. In addition, policies designed to force the acquisition of U.S.–based IPR to help combat climate change threaten U.S. export potential and the efficient diffusion of climate mitigation technologies.

The U.S. Government addresses trade barriers through international agreements such as bilateral and regional free trade agreements (FTAs), bilateral investment treaties, the WTO, existing trade and investment framework agreements, and other bilateral consultative meetings and dialogues with non-FTA partners. Taken together, the agreements provide mechanisms for addressing a wide variety of barriers to trade and investment, including barriers affecting goods and services related to renewable energy and energy efficiency.

The U.S. Government will continue to press for a favorable outcome in liberalizing trade in environmental goods and services, including those related to renewable energy, in the Doha Round negotiations. It will also explore fast-tracking WTO action in the organization's work on liberalizing trade in climate-friendly technologies, an important area in which trade could contribute to climate change mitigation and could provide momentum to the ongoing Doha negotiations.

The United States and the European Union have also created the U.S.-EU Energy Council, which, among other activities, is working on issues regarding standards in energy efficient buildings, equipment, and utilities. The comments received from the RE&EE industry emphasized the important contribution that addressing these barriers could make in creating new export opportunities for U.S. companies.

## Access to Financing

Access to affordable and sufficient amounts of financing is a major challenge shared by RE&EE companies, whether serving the domestic market or exporting. One major U.S. solar company indicated that the availability of low-cost capital in the form of loans and grants is a key element in influencing decisions to enter the export market. Financing difficulties for exporters are compounded by the inexperience of many foreign banks and regulators in assessing the financial viability of new forms of electricity generation and demand response such as RE&EE projects. In addition, U.S. firms often compete against foreign companies receiving significant financial support from their governments.

Whereas the U.S. Trade and Development Agency (USTDA) provides grant funding for RE&EE project development, including the funding of pilot projects to demonstrate U.S. technology in the field, governments in Europe and Asia have made larger amounts of funds and have open political support available to help their companies. Examples of more substantial support provided by foreign governments include (a) grant funding for projects that demonstrate technologies and establish their technical feasibility in host countries, (b) renewable energy resource assessments that enable a project to become "bankable," and (c) assistance with preparing procurement rules that can tilt the project selection to the donor country's technology. Japan, for example, provided ¥5.7 billion ($66 million) in 2008 to demonstrate RE&EE technologies in Asian countries.[8]

As a result, U.S. exporters often find themselves competing with companies whose governments are investing greater amounts to reduce the cost of capital and to increase the volumes of financial assistance available for exports. In fact, sometimes financing is tied directly to the actual export of a specific product or service.[9] Tied aid includes a grant element and additional concessional elements, such as longer repayment terms or lower (or below market) interest rates. Organization for Economic Cooperation and Development (OECD) countries reported $2.7 billion in tied and untied aid for renewable energy during 2005– 2009. The wind industry received the largest amount of tied aid support (about 46 percent), with hydro receiving the second-largest share (about 35 percent).[10]

## Creation of New RE&EE Markets

Even if all barriers to accessing foreign markets were eliminated, significant untapped markets for RE&EE products would remain. Many countries with vast potential RE&EE markets remain unattractive to U.S. exporters because of underdeveloped policy and regulatory frameworks. Working with foreign governments to help incentivize demand for RE&EE

products is one of the most significant tools the U.S. government can use as it seeks to boost RE&EE exports. Anecdotal evidence suggests that U.S. competitors are far more aggressive in leveraging public–private partnerships to assist foreign governments in developing robust markets for their companies.

The United States can do more to assist promising markets by expanding partnerships between U.S. businesses and foreign governments and their industry. For example, in July 2009, the Department of Commerce led a successful clean energy policy mission to Indonesia. The mission involved seven U.S. Government agencies and included two days of bilateral policy discussions, as well as a one-day industry forum. Several companies commented that similar missions help U.S. companies address barriers to trade and introduce them to key in-country decision-makers.

In isolation, those efforts, while helpful, will not substantially open foreign markets for U.S. companies. However, if taken as part of a coordinated and sustained effort across U.S. Government agencies to combine cooperation on technical and policy issues with trade missions, reverse missions, and subsequent policy outreach, the efforts can coalesce into an effective tool to open markets to U.S. RE&EE products.

# PART II: THE ACTION PLAN

The Initiative's action plan is a collection of commitments from several U.S. Government agencies. Together they create a framework to facilitate private-sector efforts to increase U.S. RE&EE exports over the next five years. As demonstrated by the competitiveness assessment, many of the challenges faced by U.S. RE&EE companies result from the lack of a strong national policy to provide incentives for developing domestic RE&EE technologies. Although the Initiative's commitments would undoubtedly be strengthened by domestic policies that place a price on carbon emissions, The Initiative's Action Plan commitments outline steps to increase support for RE&EE exports because U.S. companies compete against foreign firms in markets around the world every day. While it is important to emphasize the association between enacting comprehensive climate and energy legislation and further enhanced opportunities for RE&EE exports, the Initiative's Action Plan will be undertaken within existing budgets and with authorities already granted to the TPCC agencies.

Each commitment will be implemented to be consistent with federal law and international agreements. The Initiative is organized around four strategic priorities identified in the competitiveness assessment that represent the common export challenges facing U.S. RE&EE companies: (a) to tailor financing to RE&EE companies, (b) to improve market access, (c) to enhance information and trade promotion efforts to link buyers and sellers, and (d) to strengthen U.S. Government services.

## Tailor Financing to RE&EE Companies

As identified in Part I of the Initiative, access to financing and credit is often the most significant barrier facing U.S. companies as they seek to export RE&EE technologies. Without financing, U.S. companies cannot find foreign buyers, close deals, or transport goods in a timely manner. The first priority of the Initiative is, therefore, a commitment to expand the financing available to U.S. RE&EE companies. Agency commitments build on the significant financing programs already offered by U.S. Government agencies, including the Export–Import Bank of the United States (Ex–Im Bank), the Overseas Private Investment Corporation (OPIC), and the U.S. Trade and Development Agency (USTDA).

***Commitment 1: Create New Financing Products and Investment Vehicles for RE&EE Projects and Technologies***

The Ex–Im Bank has identified several constraints in the execution of renewable energy transactions, including a continued lack of liquidity in the market, limited access to pre-export working capital for production expansion, and the availability of foreign tied aid to support renewable energy exports in key markets. The Ex–Im Bank will review those hurdles and recommend policy solutions.

The USTDA increased its investments in clean energy projects from 23 percent of program funds in fiscal year 2009 to nearly 50 percent in fiscal year 2010. This increased focus on RE&EE projects, which USTDA will maintain in the years ahead, will allow more U.S. companies to participate in project preparation activities in developing and middle-income countries. Those activities will, in turn, increase the potential export opportunities for U.S. companies willing to supply technology or expertise to those projects.

In addition, the Ex–Im Bank and OPIC will unveil new products aimed at supporting aspects of the RE&EE industries under served by existing programs. OPIC, for instance, will undertake a process to support additional private equity funds to facilitate the investment of risk capital to companies or projects within the renewable energy and clean technology sectors. This process will supplement the $505 million in financing for the clean energy investment funds that were approved by OPIC's board in September 2008 but which have had to contend with the global financial crisis in trying to raise private capital. OPIC's new commitment responds to heightened private-sector interest in additional equity financing as the global economy continues to rebound and the demand for clean energy projects is renewed.

OPIC will also develop a financing tool to support energy efficiency investors in emerging markets. Specifically, OPIC will offer subordinated debt to finance the engineering and architectural design for the most energy efficient measures possible—and will finance 100 percent of the improvements. Similar to the financing offered by energy service companies, the project company will repay the subordinated debt through cost savings from the financed energy efficiency improvements.

To encourage customers interested in RE&EE products but deterred by high up-front capital costs, OPIC will begin to support the leasing of U.S. renewable energy and clean technology equipment. Because decisions to invest in RE&EE technologies are often based on capital considerations and payback periods, leasing will provide buyers additional opportunities to purchase U.S.–made RE&EE equipment.

Through these initiatives, U.S. trade and investment agencies are focused on supporting U.S. companies and investors willing to explore commercial opportunities in the RE&EE sectors of emerging markets that pose risks and otherwise lack access to capital.

### *Commitment 2: Streamline the Processing of RE&EE Transactions*

Financing RE&EE projects is often complicated, involving many investors and multiple sources of financing.

If U.S. companies cannot access financing from U.S. Government agencies, or if the process of obtaining credit is too lengthy, investors and potential importers of U.S. technology will turn elsewhere. To address this situation, both OPIC and Ex–Im Bank will create a streamlined procedure for processing RE&EE transactions. Ex–Im Bank recently put in place a Solar Express program for expedited processing of small solar transactions of at least $3 million but no more than $10 million. Further examples of potential streamlined procedures include a renewable energy fast track program that will allow the expedited approval of transactions involving renewable energy technologies. The speed with which OPIC and Ex–Im Bank will review proposals could reinforce the competitiveness of U.S. companies that are bidding on projects overseas.

## Improve Market Access for U.S. RE&EE Companies

U.S. companies cannot export their products or expertise if foreign markets are closed to U.S. goods and services. Unfortunately, several key potential export markets for the U.S. RE&EE industry have protectionist policies in place, which limit the current export potential of U.S. companies. The U.S. Government is committed to addressing those barriers to trade and will continue to promote free, fair, and open markets worldwide.

### *Commitment 3: Address Trade Barriers Facing U.S. RE&EE Companies in Foreign Markets*

The U.S. Government is committed to addressing barriers facing U.S. RE&EE companies. Barriers can take a number of different forms, from simple import tariffs to non-tariff barriers to domestic industry subsidies, which are often more complicated to address. Reducing trade barriers will increase U.S. export opportunities.

Addressing trade barriers is coordinated through the interagency trade policy mechanism, which includes the Trade Policy Staff Committee chaired by the Office of the U.S. Trade Representative (USTR). A subcommittee will review barriers and issues related to trade in clean energy technologies, including RE&EE technologies.

USTR will continue to address trade barriers through bilateral and regional FTAs, bilateral investment treaties, and ongoing negotiations in the WTO, as well as through existing trade and investment framework agreements and other bilateral consultative meetings and dialogues with non-FTA partners. Taken together, the agreements address a wide variety of barriers to trade and investment, including barriers affecting goods and services related to RE&EE. For example, USTR will continue to press for a favorable outcome in liberalizing trade in environmental goods and services, including goods and services related to renewable energy, in the Doha negotiations. It will also explore fast-tracking action in the WTO's work on liberalizing trade in climate-friendly technologies—an important area in which trade could contribute to mitigating climate change and provide momentum to the ongoing Doha negotiations.

***Commitment 4: Create New RE&EE Export Markets through Policy Advocacy and Technical Assistance to High-Potential Countries***

To support the Initiative, the Department of Commerce will lead multiple RE&EE trade policy missions in association with other TPCC agencies and in collaboration with the U.S. private sector. The missions will target countries that have significant potential for RE&EE deployment but lack the enabling policy and regulatory environment necessary to create market demand.

Once foreign governments create incentives for RE&EE technologies, a lack of regulatory, policy, or technical capacity can become a key export barrier. In developing economies, projects often lack transparent tendering and standard documentation for power purchase and energy service agreements to attract foreign investors. As such, USTDA will increase collaboration with developing and middle-income country governments and project sponsors to provide standard documentation that can accelerate the implementation of RE&EE projects by providing increased export opportunities for U.S. companies.

## Enhance Information and Trade Promotion Efforts to Link Buyers and Sellers

Through the Initiative, the U.S. Government will greatly expand opportunities for U.S. RE&EE companies to develop relationships with overseas partners by improving the information available to U.S. companies regarding foreign markets and to foreign buyers on U.S. technology. The Initiative will ensure that U.S. companies can demonstrate their expertise and meet with potential customers and clients.

***Commitment 5: Rate those Markets Most Likely to Result in RE&EE Exports for U.S. Companies***

U.S. Government resources devoted to facilitating RE&EE exports should target the most promising markets for U.S. companies. To that end, the Department of Commerce will develop a methodology to rate those markets with the greatest potential for U.S. RE&EE exports.

In addition, the Departments of Commerce and Energy will cooperate to expand the availability of U.S. export and import data to address the persistent lack of public- or private-sector information on renewable energy exports and imports.

To address the difficulty of identifying energy efficiency exports, the Department of Energy will commission studies to define energy efficiency products that are likely to be competitive in global markets. The study will identify the current level of competitiveness for U.S. firms in specific energy efficiency technologies and services by helping to focus U.S. Government resources on those energy efficiency subsectors most capable of supporting U.S. exports.

The Department of Agriculture will also expand its annual reports on biofuels to include analysis on biomass in the form of wood pellets and wood chips in relevant countries thereby providing U.S. industry and policy-makers with much-needed information on the sector's growth, export opportunities in emerging markets, and policy updates.

### *Commitment 6: Increase the Exposure of Foreign Buyers to U.S. RE&EE Companies and Technologies*

Trade missions and reverse trade missions can be effective tools for increasing exports, especially when multiple U.S. Government agencies are involved and the mission is focused on markets and industries with significant export potential. Therefore, the Department of Commerce will coordinate three to five TPCC-led RE&EE trade missions annually to key priority markets. Secretary of Commerce Gary Locke recently led the Administration's first cabinet-level trade missions to China and Indonesia. Both missions focused on the clean energy sector and demonstrated the Obama administration's commitment to the RE&EE sector, as well as the significant export opportunity that RE&EE technologies represent. Initial sales during the mission were reported at more than $20 million, with more expected as relationships are strengthened and business deals are negotiated.

The increase in the number of TPCC-led RE&EE trade missions will be complemented by reverse trade missions supported by the USTDA and DOC. The reverse trade missions will bring senior procurement officials to the United States to witness U.S. RE&EE technologies firsthand before final procurement decisions. The missions will be modeled on the highly successful Renewable Energy in Latin America Initiative organized by USTDA in 2010 that brought foreign delegations to large RE&EE trade shows in the United States.

In markets already supportive of RE&EE development, exposing foreign buyers to U.S. companies and U.S. technology is often the most important element of U.S. trade promotion activities. Therefore, during the next 12 months, the Department of Energy will work with U.S. RE&EE trade associations to assess the potential for creating foreign buyers guides for U.S. RE&EE technologies.

In addition, the TPCC Working Group will use the requirements of Executive Order 13514, “Federal Leadership in Environmental, Energy, and Economic Performance,” to showcase U.S. technologies at U.S. embassies in key target markets. The Department of Commerce will work with the Department of State to enlist 12 U.S. embassies for the program, which can provide U.S. companies with a means to showcase U.S. technology and, ultimately, to promote sales to foreign buyers.

The Department of State will also boost the U.S. Government’s involvement in the EcoPartnerships Program, which was established under the U.S.–China 10-Year Framework for Cooperation on Energy and Environment. The program harnesses the collective work of state and local governments, academic institutions, and private companies to share best practices, to foster innovation, and to further sustainable economic development between national and subnational leaders in the United States and China.

Finally, the Department of Agriculture will actively seek to identify an appropriate U.S. wood pellet industry partner to undertake foreign market development and export promotion activities under the Foreign Agricultural Service’s Market Access Program and Foreign Market Development programs. Wood pellets represent a significant export market opportunity for the U.S. biomass industry. Pursuing this opportunity will not only develop new export markets for U.S. firms, but will also capitalize on existing competitive advantages for U.S. exporters in a technology subsector currently overlooked by standard U.S. Government programs.

## Strengthen U.S. Government Services

Successful implementation of the Initiative depends on well-coordinated programs. Attempts to find and use services spread across multiple agencies takes resources and time that most RE&EE companies simply do not have. In addition, to be effective, U.S. Government programs must adapt to the dynamic nature of the RE&EE industries. The Initiative, therefore, includes new tools that enable businesses to access U.S. Government resources and that create mechanisms through which the government can consult regularly with the private sector on the effectiveness of existing programs.

Several companies identified improved coordination as a major first step toward increasing RE&EE exports, repeatedly citing missed opportunities and the potential to dramatically increase exports with additional support. Improving coordination will require a new commitment to share information across agencies and to gather information from industry on competitiveness issues facing U.S. companies both at home and abroad. Information sharing, however, is a reciprocal endeavor that involves both gathering information from industry and more effectively sharing information with industry.

***Commitment 7: Solicit Industry Advice to Improve Implementation of the Initiative, Address New Barriers, and Identify New Market Opportunities***

Close consultation with the U.S. RE&EE industry is essential to creating the policies and programs necessary for the Initiative to be effective. To provide this type of advice from industry, the Department of Commerce has created an advisory committee on RE&EE that can provide advice directly to the Secretary of Commerce on competitiveness issues facing the U.S. RE&EE industry in the United States and worldwide.

The advisory committee will include representatives from across the RE&EE industry and will provide the U.S. Government with firsthand knowledge of the issues and challenges facing U.S. RE&EE companies in export markets. The response to a Federal Register notice inviting nominations to the committee was three times the number of available appointments.[11] The first meeting will take place in December 2010.

***Commitment 8: Provide U.S. Companies with Better Information on U.S. Government Export Programs***

Several RE&EE companies requested a single guide to U.S. programs that facilitate exports. In response, the Department of Commerce will develop and maintain a Web portal hosted on www.export.gov that will include information on all U.S. Government trade facilitation programs for U.S. RE&EE exporters. The web site will include a comprehensive online guide to U.S. Government programs, market information, recent news, and a list of upcoming trade promotion events. In addition, the Department of Commerce will develop a brochure to be distributed at trade events, which will highlight the Web portal and provide information on U.S. Government services.

The lack of a one-stop shop for U.S. companies on RE&EE trade promotion services and trade events has been exacerbated by the lack of a single U.S. Government presence at RE&EE trade shows. Currently, each agency purchases separate booth space at trade shows, which is not conducive to cooperation between government agencies. This method is a wasteful and uncoordinated approach to communicating with potential exporters. In the future, the U.S. Government will operate a single unified trade show presence that will provide potential exporters with more complete information on U.S. Government programs, all while reducing costs for TPCC agencies.

## SUMMARY AND CONCLUSION

The approach outlined in this report is intended to take the first steps toward boosting and coordinating government resources and activities on behalf of RE&EE exporters. By taking into account comments from stakeholders, by reviewing existing government programs, and by thinking creatively, the action items outlined in this document constitute a significant shift in priorities and attention to U.S. RE&EE companies.

The collective effect of the action items will increase U.S. companies' ability to participate in a global market for RE&EE goods and related services, as well as offer a significant boost in U.S. Government support for the job-creating potential of the RE&EE industry.

The Initiative was designed to be implemented under existing budgets. However, if several of the action items are to be sustained, additional funding and continued engagement will be required. In the near term, some commitments contained in this Initiative can help immediately by informing U.S. RE&EE exporters of the resources available and by coordinating these resources and agencies more effectively. In the midterm, the Initiative aims to fulfill the NEI's goals established by President Obama. In the long term, the Initiative's ultimate success will be measured by the ability of U.S. companies to compete in overseas markets.

Through regular meetings and evaluations, as well as through feedback from the RE&EE advisory committee and other stakeholders, the TPCC Working Group on RE&EE will consider new ideas that arise while other action items reach their goals or evolve into new initiatives.

The pace of innovation in the RE&EE sector is significant and will likely lead the RE&EE industries in unimaginable directions during the next five years. The structure of the Initiative is flexible as different needs and opportunities arise, but it is firm in its commitment to supporting U.S. RE&EE companies as the U.S. Government works toward a fair and unbiased international trading arena.

**Box 4. 23 Deliverables Across**
**8 U.S. Government Agencies**

Eight TPCC agencies have committed to new actions or programs that will help meet the goals of the RE&EE Export Initiative. Each deliverable will be implemented within existing budgets and within existing authorities already granted to TPCC agencies. A list of the RE&EE Export Initiative's deliverables is below:

*Tailored Financing for RE&EE Companies*

- Streamlined agency processing of RE&EE transactions at both Ex-Im and OPIC
- Energy efficiency subordinated debt product from OPIC
- OPIC commitment to supplement the $505 million in financing for six clean energy investment funds announced in 2008
- Support for U.S. leasing sector of RE&EE equipment
- Increased USTDA funding for RE&EE projects through additional reverse trade missions and pre-feasibility studies

*Improved Market Access for U.S. RE&EE Companies*

- TPCC-led renewable energy and energy efficiency trade policy missions organized by the Department of Commerce
- Expansion of USDA's Foreign Agricultural Service's Market Development Program to include wood pellets
- Development of standard RE&EE technical documentation through USTDA to expedite RE&EE projects in developing and emerging markets
- Creation of a TPSC subcommittee focused on RE&EE by the Office of the U.S. Trade Representative

*Enhanced Information and Trade Promotion Efforts to Link Buyers and Sellers of U.S. RE&EE Products and Services*

- Department of Commerce study rating top RE&EE markets for potential U.S. exports

- Foreign buyers guides to U.S. RE&EE technology developed by the Departments of Commerce and Energy
- Increased number of TPCC-designated RE&EE trade missions
- Department of Energy study on the export potential of energy efficiency products and services
- Expansion of the Green Embassies Program
- Unified U.S. Government outreach at key RE&EE trade shows
- Boosting exports through EcoPartnerships Program

*Strengthen U.S. Government Services for U.S. RE&EE Companies*

- Creation of a RE&EE Advisory Committee to provide private-sector expertise to the Secretary of Commerce
- TPCC RE&EE web portal on export.gov
- Online exporters guide for U.S. RE&EE companies
- Development of a U.S. Government RE&EE exporters brochure
- OPIC organizational focus on RE&EE priorities

# ANNEX 1: DEFINITIONS

The Renewable Energy and Energy Efficiency Export Initiative (Initiative) focuses on equipment and services related to electricity production from renewable energy sources, as well as to the products and services used to promote energy efficiency improvements. Rapid technological advances in the design, manufacturing, and application of RE&EE equipment make comprehensive definitions of these sectors difficult. The following definitions should be considered illustrative rather than definitive. They are intended to be used specifically for measuring progress toward meeting the Initiative's goals and not for other assessments regarding the RE&EE industries. General agreements about which industry sectors are included within the scope of the Initiative are essential for benchmarking current exports and for assessing the competitive state of the industry. For a list of product codes used in calculating exports in each subsector, please refer to Annex 3.

## Renewable Energy

Equipment and services related to all types of electricity generation from renewable sources is included within the scope of the Initiative. Main types of renewable energy are listed next.

*Biomass:* Biomass is biological material derived from living, or recently living, organisms. In the context of the Initiative, biomass-based electricity can be generated either from plant-based material or solid waste or from landfill gas. Stationary biomass power plants burn predominantly products that would otherwise crowd landfills or pose a fire hazard in forests.[12] Common fuels include agricultural residues, forest underbrush, urban waste, landfill gas, municipal solid waste, and high-yield energy crops grown specifically for biomass use. Biomass exports include wood pellets, which are easier to transport and easier to co-fire with existing coal power plants.

*Note: The Initiative does not include the clean transportation sector and, therefore, does not consider exports of electric vehicles, biofuels, or other transportation-related technology. The exclusions are for scoping purposes only and should not be construed as value judgments.*

*Geothermal:* Geothermal technologies use heat from the earth to provide heating and cooling for homes and buildings or for electricity generation through steam. Geothermal heat is captured in a variety of ways:

- Hydrothermal collects steam from hot water reserves underground.
- Low-temperature binary units often coproduce with existing oil and gas infrastructure.
- Enhanced geothermal fractures dry rock and pumps water into the fracture to produce steam.
- Direct use produces heat directly from hot water found underground.
- Heat pumps use temperature differences in the shallow ground to heat and cool buildings.

Hydropower: Hydropower uses the kinetic energy from running water to drive an electricity-generating turbine. For the purposes of the Initiative, hydropower includes equipment for the conventional hydropower industry, pump storage, as well as small and micro hydro products and in-stream hydrokinetics. Hydropower also includes ocean energy technologies that use the temperature differences within the ocean or the kinetic energy from tides and waves to produce electricity.

Solar: A variety of technologies have been developed to take advantage of energy generated by the sun, including the following:

- Concentrated Solar Power (CSP). Also called solar thermal power, it uses the sun as a heat source to produce steam, which spins a large turbine that drives an electricity generator.
- Photovoltaics (PV). Solar cells are made from semiconducting materials that convert sunlight directly into electricity.
- Solar hot water. Systems can be either active or passive, but most are active systems that rely on pumps to move liquid between the collector and the storage tank. Passive systems rely on gravity and natural circulation.

Wind: Wind turbines are structures that convert kinetic wind energy into electricity. The majority of wind turbines are deployed in large, utility-scale groupings, commonly known as wind farms. Smaller wind turbines provide power to facilities such as skyscrapers, factories, or farms. A megawatt-scale wind turbine is a massive structure with a rotor diameter that frequently exceeds 100 meters. A typical wind turbine consists of a tower, a nacelle, blades, and about 8,000 additional subcomponents.

## Energy Efficiency

The Initiative seeks to boost the competitiveness of goods or services that are specifically designed to use less energy to perform the same function, to improve performance with the same energy inputs, or to incorporate previously wasted by-products to reduce overall energy use. Examples of energy efficiency goods and services include green building design services and materials; industrial energy efficiency applications such as combined heat and power; district energy systems; and, where clear standards can be identified, appliances and electronics designed and marketed as energy efficient.

Two broad delineations in energy efficiency are energy *efficient* products and energy *efficiency* products. Energy efficient products are those products that use less energy than similar goods, and a standard is often set as a benchmark for energy use. For instance, a compact fluorescent light bulb is an energy efficient product compared with an incandescent light bulb. Energy efficiency products are those products that help people use energy more efficiently, such as a programmable thermostat or motion sensor lighting.

Industry stakeholders also expressed that efficiency has both demand- and supply-side implications. The term energy efficiency can be narrowly associated with conservation or demand-side efficiency (that is, reducing the amount of electricity consumed by end users, such as households, commercial buildings, and industrial processes). Supply-side energy efficiency (that is, reducing the amount of fuel used to generate electricity or heat) is a piece of the efficiency spectrum that is often overlooked. Supply-side energy efficiency alternatives for electricity generation might include replacing a coal-fired boiler process with a gas-fired turbine or a combined heat and power process or upgrading the transmission system to reduce losses. Given those considerations, benchmarking U.S. exports of energy efficiency products requires further definition to provide an accurate estimate.

The development of methodologies to track the export potential and trade flows of U.S. energy efficiency products will enable a greater focus on the competitiveness of specific products. Until the data can be improved, U.S. trade facilitation efforts are focused on categories of goods and services generally associated with the energy efficiency industry, including energy efficiency buildings; industrial applications, such as combined heat and power; demand-control technologies associated with the smart grid; and some appliance, electronic, and communications products specifically designed and marketed to meet energy efficiency criteria.

## ANNEX 2: BENCHMARKING U.S. RE&EE EXPORTS

The United States exported at least $2 billion worth of renewable energy products in 2009. Wind and solar technologies account for a vast majority of renewable energy exports (85 percent). The export of services, such as design, consultancy, or engineering—which are often the most significant export markets for the U.S. hydropower, geothermal, and biomass industries—are not reflected in this report.[13]

### Wind

The United States exported $318.7 million of wind energy equipment in 2009, with an annual average growth of 29.1 percent between 2007 and 2009.

Imports dwarfed exports, however, with imports of more than $3.67 billion in 2009. Average annual imports grew slightly from 2007 to 2009 at an average rate of just under 0.35 percent annually. The trade imbalance is because of the rapid rise of U.S. demand for wind turbines, which has outstripped U.S. wind energy manufacturing capacity.

The outlook for U.S. wind energy manufacturing is slowly improving. Incentives and increased demand have resulted in investment in U.S. production facilities. The average wind turbine installed in the United States contains between 50 percent and 60 percent U.S.–made content. Imports are also strong for blade and wind energy components despite a surplus of U.S. manufacturing capacity. U.S. blades and towers face fierce competition from foreign imports because of differences in transportation costs. Blades and towers are easy to ship but costly to transport by truck or rail.

The development of a U.S. offshore wind market could result in significant manufacturing investments in coastal states. Although initial investments would service the U.S. market, the related nature of the wind supply chain could encourage future exports. Market analysts predict that a significant U.S. offshore wind market is unlikely before 2014 because of difficulties with permitting and financing.

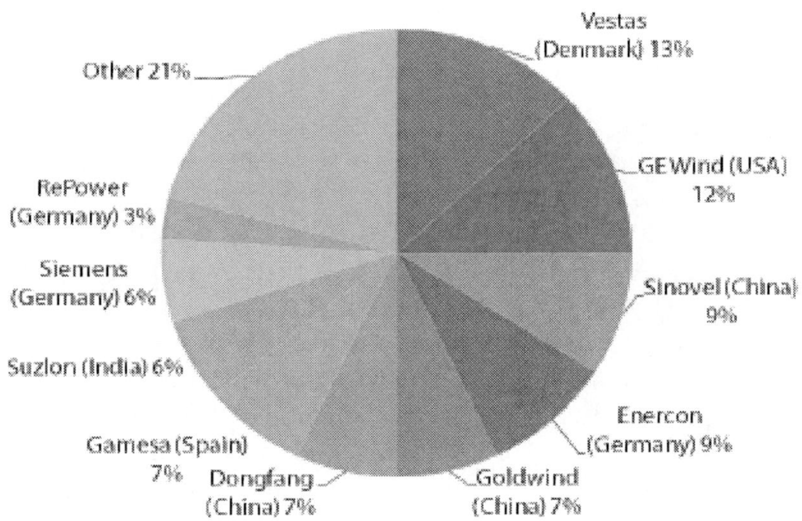


Source: Renewable Energy Policy Network for the 21st Century, "Renewables 2010 Global Status Report."

Figure 2. Wind Turbine Manufacturers, By Market Share (2009).

Lack of data on the export of services for the global wind industry leaves information on U.S. wind exports incomplete. Anecdotal evidence indicates that the value of service exports, such as licensing, siting, project development, and consulting, is substantial. U.S. companies have been successful licensing wind turbine designs to foreign manufacturers. Licensing agreements are also often coupled with service contracts for components, which can provide additional export revenue.

The small wind industry also offers the potential for strong exports. The global small wind market, defined as the market for wind turbines with a generating capacity of 100 kilowatts or under, was estimated at $156 million in 2008 and is responsible for more than 37 megawatts of wind capacity. Although still relatively nascent, the global industry grew by 53 percent during 2007–2008. The small wind sector has also defied the financial crisis to grow rapidly between 2008 and 2010. Most important, U.S. manufacturers accounted for approximately 49 percent of global sales, of which 29 percent were exports.[14]

## Solar

### *Photovoltaics (PV)*

Exporters of U.S.–manufactured PV equipment face a highly competitive marketplace (see Figure A2.2).

Despite capturing only 6 percent of global manufacturing capacity, the United States exported a relatively robust $1.18 billion of cells, modules, and panels in 2009.

The solar PV supply chain includes polysilicon feedstock, wafers, cells, balance of systems, crystalline silicon (c-Si) modules, thin-film modules, and concentrated PV modules. The United States is a leading producer of polysilicon feedstock and its production technology.

Polysilicon is the primary feedstock of c-Si modules. U.S. companies also have a strong competitive advantage in innovative thin-film modules, which account for 10 percent of PV modules shipped or installed globally.[15] Although demand is unpredictable, U.S. manufacturers of concentrated PV modules appear to be highly competitive. U.S. companies captured 69 percent market share of concentrated PV capacity in operation or soon to be online.[16]

U.S.–based c-Si companies face fierce competition. Because of widespread use of glass, c-Si modules are more costly to ship long distances than thin-film panels. Chinese companies currently account for 40 percent of the c-Si market. Prices of the c-Si module have dropped significantly since 2008, putting pressure on company margins. Although better for consumers, lower panel prices and strains on national budgets are prompting a number of European countries to reevaluate the levels of financial incentives for solar PV. The policy changes create a degree of demand uncertainty for all PV technologies. U.S.–based manufacturers that can innovate low-cost manufacturing techniques and identify outlets for their products, either by building relationships with developers or by initiating projects themselves, will be best placed to successfully compete.

***Solar Water Heating***

Solar water heating (SWH) is a subcategory of solar thermal technology, which collects the sun's energy to heat water. The technology represents small- and medium-sized solar collectors, which are generally used in domestic water, space, and swimming pool heating. In 2009, China manufactured approximately 77 percent of all solar thermal collectors and had 70 percent of global installed capacity.[18] U.S.–based manufacturers continue to find markets. SWH exports grew slightly to $138.7 million in 2009, up from $128 million in 2008. By contrast, the United States imported a mere $33.1 million in 2009, a decrease of approximately 50 percent from 2008.

***Concentrated Solar Power***

The United States, along with Spain and North Africa, are expected to be the dominant markets for concentrated solar power (CSP) technology. CSP projects are in operation only in the United States and Spain, although many more projects are expected to be completed within the next five years. European manufacturers dominate the equipment market for receivers and mirrors. Accordingly, although the global CSP market is still small, U.S. imports are significantly higher than exports. Some European manufacturers have established manufacturing plants in the United States, which will likely be used for projects in the United States. U.S. exports in this category are centered on electric generators used to generate electricity from the steam produced by the CSP plant. Development services and operations expertise are another area of U.S. competitive strength that should help to generate additional exports. New innovations could also place U.S. manufacturers in a position to capture market share from global competitors.

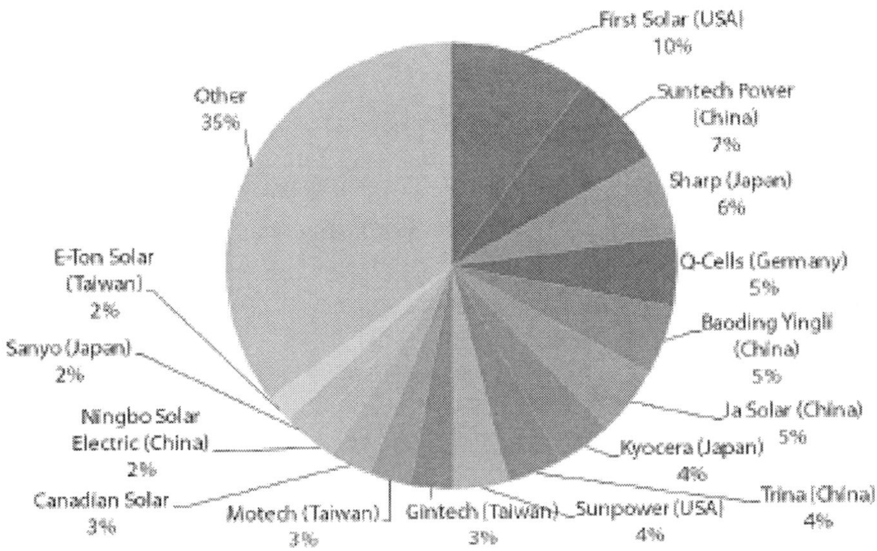


Source: Renewable Energy Policy Network for the 21st Century, "Renewables 2010 Global Status Report."

Figure 3. Solar PV Manufacturers, By Market Share.

## Geothermal

The United States exported $70.1 million worth of geothermal equipment in 2009. Over time, U.S. companies have developed the capacity, expertise, and ability to supply technology to the U.S. market and, as a result, can often supply global markets when opportunities arise.

Export and import totals are likely much higher than the numbers suggest, given the lack of available data (the result of product differentiation in the Harmonized Tariff Schedule (HTS) codes that are used to calculate geothermal exports and imports; see Annex 3 for HTS codes). In addition, anecdotal evidence suggests that a majority of geothermal exports are service related and are, therefore, not included in this report's estimate of total geothermal exports. U.S. companies enjoy a competitive advantage in geothermal drilling, financing, design, and engineering.

As global demand for geothermal energy expands, U.S. companies are likely to be competitively positioned to take advantage of new opportunities.

In 2005, 8.9 gigawatts (GW) of total geothermal capacity was installed in 24 countries, according to the International Geothermal Association. By the end of 2010, 10.7 GW of total capacity was installed in 70 countries. The pipeline of new projects is even larger, with 12.3 GW currently under development, more than double the world's current installed capacity.[17]

The trend of expansion in conventional geothermal technologies is complemented by the development of new techniques that could improve the industry's growth prospects during the coming years. Several European countries that have not been involved in geothermal energy production are now using low-temperature technologies to produce geothermal energy domestically, including enhanced geothermal systems.[18] The use of geothermal heat pumps has also expanded worldwide. The heat pump industry in the United States has experienced 10 percent annual growth during the past decade and now exports to at least 28 countries.[19]

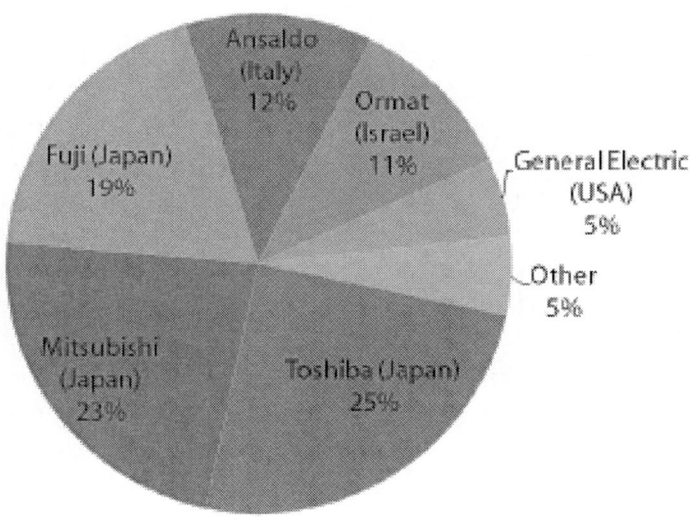


Source: Renewable Energy Policy Network for the 21st Century, "Renewables 2010 Global Status Report."

Figure 4. Geothermal Turbine Manufacturers, By Market Share (2009).

In the near term, exports from the United States are likely to increase in the subsectors that currently enjoy a competitive advantage, including the drilling, financing, and engineering sectors, as well as the growing geothermal heat pump industry. The destinations of U.S. exports are closely tied to the development of policy incentives in countries worldwide.

Countries with strong incentives for geothermal technologies and developed regulatory frameworks can experience robust growth, whereas those with weak policies are likely to experience inconsistent growth and ultimately reduced interest from companies determined to invest in regions with long-term policy certainty. Argentina, Chile, Greece, Hungary, Indonesia, the Philippines, and Turkey, as well as countries in Central America, combine both potential geothermal capacity and a strong policy environment; therefore, U.S. exports to those countries could continue to increase.

## Biomass

The United States exported $176.4 million in biomass energy equipment and feedstock in 2009, with an annual average growth of 54 percent between 2007 and 2009. Imports during the same period were $349.2 million worth of biomass equipment, with an average annual import growth of 28 percent.

Several countries are expanding their use of biomass for power, either by building biomass-specific power plants or by co-firing biomass in existing coal-based power plants. Many European countries already obtain a substantial portion of their electricity from biomass, most notably Sweden, which produced more energy from biomass than from oil in 2009.[20] Several developing countries have recently developed biomass power capacity, including Brazil, Costa Rica, India, and Mexico, but developing countries that consume biomass resources often use their own domestic resources rather than import feedstock from the United States.

The United States currently exports wood pellets and wood chips to Europe for co-firing in existing coal plants. As more countries enact carbon reduction requirements, co-firing could become increasingly common. In addition, future exports in the biomass industry are likely to be in the form of consulting, engineering, procurement, and financial services, all industries in which the United States is likely to remain competitive. U.S. companies should find relevant export opportunities in countries with little available feedstock or without a local biomass industry, particularly if strong government policy in those countries supports the use of biomass for power.

## Hydropower

The United States exported $67.7 million of hydroelectric energy equipment in 2009, with an average annual growth of 44.56 percent between 2007 and 2009. The United States maintains a positive trade balance in the sector, with imports totalling only $57.4 million in 2009.

The global growth in the hydropower industry is driven primarily by the technology's low generation cost relative to other renewable technologies. In addition, hydropower can produce base load electricity, reducing the need for fuel imports. In the developing world, large hydropower facilities are being constructed, often as part of programs to promote water resource management that can reduce the likelihood of floods. In countries with sizable, or growing, renewable energy production, hydropower is increasingly being used for power generation and energy storage through the use of pump-storage facilities. In addition, new technologies, such as in-stream hydrokinetics and ocean energy systems, are providing potential growth markets in coastal regions worldwide.

Many developing countries continue to build conventional hydropower capacity, with several new projects under development in Brazil, China, India, Malaysia, Russia, Turkey, and Vietnam.[20] In total, the hydropower industry added 31 GW of new capacity in 2009—the second-highest amount among renewable energy technologies (wind power saw the most growth in 2009).[21] In OECD countries, the hydropower industry is also experiencing growth, with new capacity being added through efficiency improvements at existing dams and through power production equipment on existing dams that were previously used for flood control.

Though not fully commercialized, ocean energy technologies, which capture energy from the movement of ocean waves or from temperature differences at various ocean depths, may provide additional hydropower exports by 2015. Most of the industry's growth is taking place in Europe, particularly in the Portugal, Spain, and the United Kingdom, but growth is possible in all coastal markets. An estimated 6 MW of ocean energy is currently operational or being tested in those markets.

## Energy Efficiency

The market potential for U.S. energy-efficient technologies and services is likely substantial, although specific trade data to support these projections are not available. Trade flow data on products that are considered energy efficient are not captured in the Harmonized Tariff System or North American Industry Classification System. In addition, trade data collection does not differentiate energy-efficient and conventional building and industrial products.[22] In the absence of such energy-efficient trade-related data, other indicators, such as the price of energy, energy consumption, and energy production, are used in this report to assess the potential for consumption of energy-efficient goods and services.

Specific markets for industrial energy efficiency equipment and services are primarily located in economically developed countries. Other considerations are whether the country consumes more energy than it produces domestically, such as Germany, or whether the country has a high average cost of energy, such as Japan. These factors are part of the equation for understanding the demand for energy efficiency products and should be considered when making market assessments.

The largest U.S. export market for building materials is Canada, which imports more building materials from the United States than the next 20 export markets combined. Other top export markets include Australia, China, Germany, Japan, Mexico, and the United Kingdom.

Top U.S. export markets for electronics, appliances, and information and communication technologies are Canada, China, Japan, Mexico, and Singapore, with Canada and Mexico representing the major importers.

U.S. companies are also leaders in developing and distributing emerging energy efficiency technologies, such as smart-grid equipment and services. Both industries are relatively nascent, but as demand increases, U.S. companies will be well positioned to supply these technologies to the global market. Both technologies also stand to buttress the growing global demand for renewable energy technologies, because they can provide an important counterbalance to the intermittency of solar- and wind-generated electricity. Smart-grid technologies, for instance, can be used in a variety of applications, including integrated communications technologies, electricity sensing and measuring equipment, advanced components that monitor electricity movement, control methods that enable instantaneous modifications to electricity distribution, and improved support systems that allow instantaneous input from grid operators and electricity consumers.

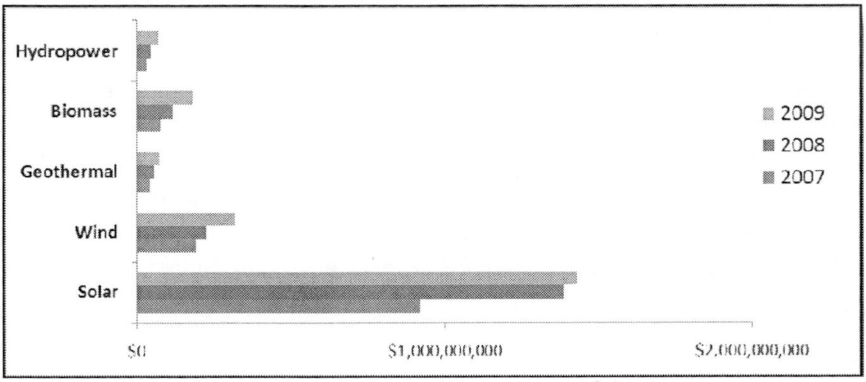


Source: International Trade Commission and Department of Commerce.

Figure 5. Renewable Energy Exports, By Sector.

# ANNEX 3: METHODOLOGY

Accurately measuring trade in RE&EE goods and services is difficult because of a lack of standard product definitions and the inability to differentiate component parts used in RE&EE applications from components used in other energy sectors or other industries. Renewable energy exports, which are calculated in the Initiative, therefore refer to those products that are identifiable as used only for renewable energy generation. The total of renewable energy exports is, therefore, likely much higher but is impossible to accurately calculate at this time.

Anecdotal evidence suggests, however, that U.S. companies are world leaders in the types of services used by the RE&EE industry, particularly the design, engineering, and financing of RE&EE projects. It is likely that export totals for the RE&EE services are substantial. Until standard product definitions are developed that separate the energy efficiency component of a product's value from its overall value, calculating the total value of energy efficiency exports will be impossible. Without standards for classifying energy efficiency products, any grouping of products considered energy efficient would be made on a one-off basis that lacked peer review. Furthermore, the export value of RE&EE services cannot be accurately calculated using existing data sources.

Therefore, to estimate renewable energy exports, the Initiative used data compiled by the Department of Commerce and a report by Senator Ron Wyden. Published in May 2010, the Wyden Report is the most comprehensive effort to date to estimate the real value of environmental goods, a subset of which are RE&EE goods.

The Department of Commerce used the codes from the Harmonized Commodity Description and Coding System identified in the Wyden report and then added more codes after consultations with industry and sector specialists (see Table A3.1).

| Wind | | |
|---|---|---|
| Product | HTS6 | HTS6 Product Description |
| Wind turbine towers | 730820 | Towers and lattice masts |
| Wind turbine rotors, shafts, and parts | 841290 | Other engines and motors and parts thereof |
| Fixed ratio speed changers | 848340 | Gears and gearing, other than toothed wheels, chain sprockets, and other transmission elements presented separately; ball or roller screws; gear boxes and other speed changers, including torque converters |
| Multiple and variable ratio speed changers | 848340 | Gears and gearing, other than toothed wheels, chain sprockets, and other transmission elements presented separately; ball or roller screws; gear boxes and other speed changers, including torque converters |
| Wind turbine couplings | 848360 | Clutches and shaft couplings, including universal joints |
| AC generators (alternators), of an output not exceeding 75 kilo volt amperes (kVA) | 850161 | AC generators (alternators), of an output not exceeding 75 kVA |
| Solar | | |
| Solar preheating storage tank | 730900 | Reservoirs, tanks, vats, and similar containers for any material (other than compressed or liquefied gas), of iron or steel, of a capacity exceeding 300 liters, whether or not lined or heat insulated, but not fitted with mechanical or thermal equipment |
| Portable solar cookers and ovens (made of steel) | 732119 | Cooking appliances and plate warmers: other, including appliances for solid fuel |
| Solar cookers and ovens (made of cast iron) | 732119 | Cooking appliances and plate warmers: other, including appliances for solid fuel |
| Solar cookers and ovens (other) | 732119 | Cooking appliances and plate warmers: other, including appliances for solid fuel |
| Parts of solar cookers and ovens | 732190 | Stoves, ranges, grates, cookers (including those with subsidiary boilers for central heating), barbecues, braziers, gas rings, plate warmers, and similar non-electric domestic appliances, and parts thereof, of iron or steel: parts |

| Product | HTS6 | HTS6 Product Description |
|---|---|---|
| Portable solar cookers and ovens (made of aluminum) | 761699 | Other articles of aluminum, not elsewhere specified or identified in chapter 76 |
| Solar water heaters | 841919 | Instantaneous or storage water heaters, non-electric: other |
| Brazed aluminum heat exchangers | 841950 | Heat exchange units, whether or not electrically heated |
| Brazed aluminum heat exchanger parts | 841990 | Parts of machinery, plant, and equipment of heading 8419 |
| Solar panels and modules | 850131 | Other DC motors; DC generators: of an output not exceeding 750 watts |
| AC generators (alternators), of an output not exceeding 75 kVA | 850161 | AC generators (alternators), of an output not exceeding 75 kVA |
| AC generators (alternators), of an output exceeding 75 kVA but not exceeding 375 kVA | 850162 | AC generators (alternators), of an output exceeding 75 kVA but not exceeding 375 kVA |
| AC generators (alternators), of an output exceeding 375 kVA but not exceeding 750 kVA | 850163 | AC generators (alternators), of an output exceeding 375 kVA but not exceeding 750 kVA |
| AC generators (alternators), of an output exceeding 750 kVA but not exceeding 10,000 kVA | 850164 | AC generators (alternators), of an output exceeding 750 kVA but not exceeding 10,000 kVA |
| AC generators (alternators), of an output exceeding 10,000 kVA but not exceeding 40,000 kVA | 850164 | AC generators (alternators), of an output exceeding 10,000 kVA but not exceeding 40,000 kVA |
| AC generators (alternators), of an output exceeding 40,000 kVA | 850164 | AC generators (alternators), of an output exceeding 40,000 kVA |
| Portable solar power generation equipment | 850239 | Other electric generating sets: other |
| Solar power electric generating sets | 850239 | Other electric generating sets: other |
| Parts for portable solar power generation equipment and electric generating sets | 850300 | Parts suitable for use solely or principally with the machines of heading 8501 or 8502 |
| Solar inverters | 850440 | Static converters |
| PV charge controllers | 850440 | Static converters |
| Deep discharge solar batteries | 850720 | Lead-acid storage batteries nesoi |
| PV system control devices | 853710 | Boards, panels, consoles, etc. with electrical apparatus, for electric control or distribution of electricity, for a voltage not exceeding 1,000 V. |
| Solar modules and panels | 854140 | Photosensitive semiconductor devices, including photovoltaic cells whether or not assembled in modules or made up into panels; light emitting diodes |
| Solar cells | 854140 | Photosensitive semiconductor devices, including photovoltaic cells whether or not assembled in modules or made up into panels; light emitting diodes |

| Product | HTS6 | HTS6 Product Description |
|---|---|---|
| Fresnel mirrors | 900190 | Optical fibers and optical fiber bundles; optical fiber cables other than those of heading 8544; sheets and plates of polarizing material; lenses (including contact lenses), prisms, mirrors, and other optical elements, of any material, unmounted, other than such elements of glass not optically worked: other |
| Fresnel reflector modules | 900290 | Lenses, prisms, mirrors, and other optical elements, of any material, mounted, being parts of or fittings for instruments or apparatus, other than such elements of glass not optically worked; parts and accessories thereof: other |
| Heliostats | 903289 | Automatic regulating or controlling instruments and apparatus; parts and accessories thereof: other |
| Temperature sensors and controllers for solar | 903289 | Automatic regulating or controlling instruments and apparatus; parts |
| boilers and water heaters | | and accessories thereof: other |
| Heliostat parts | 903290 | Automatic regulating or controlling instruments and apparatus; parts and accessories thereof: parts and accessories |
| Geothermal | | |
| Steam turbines of an output exceeding 40 megawatts (MW) | 840681 | Steam turbines and other vapor turbines, and parts thereof: of an output exceeding 40 MW |
| Steam turbines of an output not exceeding 7,460 kilowatts (kW) | 840682 | Steam turbines and other vapor turbines, and parts thereof: of an output not exceeding 7,460 kW |
| Steam turbines of an output exceeding 7,460 kW, but not exceeding 40 MW | 840682 | Steam turbines and other vapor turbines, and parts thereof: of an output exceeding 7,460 kW, but not exceeding 40 MW |
| Rotors for geothermal steam turbines | 840690 | Parts of steam turbines |
| Other rotors for geothermal steam turbines | 840690 | Parts of steam turbines |
| Blades for geothermal steam turbines | 840690 | Parts of steam turbines |
| Other parts for geothermal steam turbines | 840690 | Parts of steam turbines |
| Self-contained reversing valves of an output not exceeding 17.58 kW per hour | 841581 | Air-conditioning machines incorporating a refrigerating unit and a valve for reversal of the cooling and heating cycle (reversible heat pumps) |
| Self-contained reversing valves of an output exceeding 17.58 kW per hour | 841581 | Air-conditioning machines incorporating a refrigerating unit and a valve for reversal of the cooling and heating cycle (reversible heat pumps) |

| Product | HTS6 | HTS6 Product Description |
|---|---|---|
| Other reversing valves | 841581 | Air-conditioning machines incorporating a refrigerating unit and a valve for reversal of the cooling and heating cycle (reversible heat pumps) |
| Ground source heat pumps | 841861 | Heat pumps, other than the air-conditioning machines of heading 8415 |
| Brazed aluminum plate-fin ground-coupled heat exchangers | 841950 | Heat exchange units, whether or not electrically heated |
| Other ground-coupled heat exchangers | 841950 | Heat exchange units, whether or not electrically heated |
| Biomass | | |
| Wood waste shavings | 440130 | Sawdust and wood waste and scrap, whether or not agglomerated in logs, briquettes, pellets, or similar forms |
| Wood waste pellets and other | 440130 | Sawdust and wood waste and scrap, whether or not agglomerated in logs, briquettes, pellets, or similar forms |
| Anaerobic digester tanks | 730900 | Reservoirs, tanks, vats, and similar containers for any material (other than compressed or liquefied gas), of iron or steel, of a capacity exceeding 300 liters, whether or not lined or heat insulated, but not fitted with mechanical or thermal equipment |
| Biomass boilers | 840219 | Other vapor-generating boilers, including hybrid boilers |
| Biomass boiler heat exchangers | 840290 | Steam or other vapor-generating boilers (other than central heating hot water boilers capable also of producing low-pressure steam); super-heated water boilers; parts |
| Biomass boiler parts | 840290 | Steam or other vapor generating boilers (other than central heating hot water boilers capable also of producing low pressure steam); super-heated water boilers; parts |
| Auxiliary plant economizers for biomass boilers | 840410 | Auxiliary plant for use with boilers of heading 8402 or 8403 |
| Auxiliary plant for biomass boilers | 840410 | Auxiliary plant for use with boilers of heading 8402 or 8403 |
| Parts for auxiliary plants for biomass boilers | 840490 | Auxiliary plant for use with boilers of heading 8402 or 8403 (for example, economizers, super-heaters, soot removers, gas recoverers); condensers for steam or other vapor power units; parts |

| Product | HTS6 | HTS6 Product Description |
|---|---|---|
| Gas turbines for biofuels not exceeding 5,000 kW | 841181 | Other gas turbines of a power not exceeding 5,000 kW |
| Gas turbines for biofuels exceeding 5,000 kW | 841182 | Other gas turbines of a power exceeding 5,000 kW |
| Biogas refinement equipment | 841940 | Distilling or rectifying plant |
| Anaerobic biogas reactors, digestion tanks, and biogas refinement equipment | 841989 | Machinery, plant, or laboratory equipment, whether or not electrically heated (excluding furnaces, ovens, and other equipment of heading 8514), for the treatment of materials by a process involving a change of temperature such as heating, cooking, roasting, distilling, rectifying, sterilizing, pasteurizing, steaming, drying, evaporating, vaporizing, condensing, or cooling, other than machinery or plant of a kind used for domestic purposes; instantaneous or storage water heaters, nonelectric |
| Parts of heat exchange units for biogas reactors, digestion tanks, and biogas refinement equipment | 841990 | Parts of machinery, plant, and equipment of heading 8419 |
| Other parts for biogas reactors, digestion tanks, and biogas refinement equipment | 841990 | Parts of machinery, plant, and equipment of heading 8419 |
| Anaerobic digester equipment | 847989 | Machines and mechanical appliances having individual functions, not specified or included elsewhere in this chapter; parts thereof: other |
| Anaerobic digester parts | 847990 | Machines and mechanical appliances having individual functions, not specified or included elsewhere in this chapter; parts thereof: parts |
| Biomass combined heat and power systems and parts | 850239 | Other electric generating sets: other |
| Biomass combined heat and power systems and parts | 850300 | Parts suitable for use solely or principally with the machines of heading 8501 or 8502 |
| Hydropower | | |
| Small hydraulic turbines | 841011 | Hydraulic turbines and water wheels: of a power not exceeding 1,000 kW |
| Medium hydraulic turbines | 841012 | Hydraulic turbines and water wheels: of a power exceeding 1,000 kW but not exceeding 10,000 kW |
| Large hydraulic turbines | 841013 | Hydraulic turbines and water wheels: of a power exceeding 10,000 kW |
| Hydraulic turbine parts | 841090 | Hydraulic turbines, water wheels, and regulators therefore; parts, including regulators |

# ANNEX 4: OVERVIEW OF CURRENT ACTIVITIES AND POLICIES OF THE TPCC AGENCIES RELATED TO RENEWABLE ENERGY AND ENERGY EFFICIENCY

Below is an overview of the TPCC agencies' role in international clean energy policy, capacity-building, and export-related activities. The services available to U.S. exporters of RE&EE are described in detail in a new online guide on the TPCC RE&EE Web portal at *www.export.gov/reee.*

| Agency | Description |
|---|---|
| | The Export–Import Bank of the United States (Ex–Im Bank) is the official export credit agency of the United States. Ex–Im Bank assists in financing the export of U.S. goods and services to international markets by providing export financing products that fill gaps in trade financing and assuming credit and country risks that the private sector is unable or unwilling to accept. Ex–Im Bank has special provisions aimed at leveling the playing field for U.S. exporters by matching the financing that other governments provide to their exporters. Ex–Im Bank provides working capital guarantees (pre-export financing), export credit insurance, loan guarantees, direct loans (buyer financing), and project and structured finance. The bank works in more than 150 countries and operates on a self-sustaining basis. |
| Export–Import Bank of the United States (Ex-Im Bank) | Ex–Im Bank programs that provide special support for RE&EE include the following:<br><br>• Solar Express Initiative. Launched in April 2010, Solar Express provides streamlined project financing to small solar-power producers that meet Ex–Im Bank's credit standards. Under the initiative, the bank will consider project financing for small solar-power producers seeking loans of at least $3 million but no more than $10 million.<br>• Carbon Policy. Approved in November 2009, Ex–Im Bank's new Carbon Policy includes the establishment of a $250 million renewable energy facility, the adoption of new measures to support exports that contribute to substantial increases in energy efficiency, and a commitment to further improve the tracking and reporting of carbon dioxide emissions from Ex–Im Bank supported projects. |
| Overseas Private Investment Corporation (OPIC) | The Overseas Private Investment Corporation (OPIC) helps U.S. businesses invest overseas, fosters economic development in new and emerging markets, complements the private sector in managing risks associated with foreign direct investment, and supports U.S. foreign policy. OPIC financing provides medium- to long-term funding through direct loans and loan guarantees to eligible investment projects in developing countries and emerging markets. By complementing the private sector, OPIC can provide financing in countries where conventional financial institutions often are reluctant or unable to lend on such a basis. |

| Agency | Description |
|---|---|
| | OPIC insurance is available to U.S. investors, contractors, exporters, and financial institutions involved in international transactions. Political risk insurance can cover currency inconvertibility, expropriation, and political violence, and it is available for investments in new ventures, expansions of existing enterprises, privatizations, and acquisitions with positive developmental benefits. OPIC works in more than 150 emerging and transitioning economies and operates on a self-sustaining basis by supporting commercially viable projects on commercial terms. |
| | Renewable energy and clean energy project funding are priority areas for OPIC. Activities include the following:<br>• A dedicated renewable energy and sustainable development finance team, which offers innovative financing structures for energy efficiency and leasing of U.S. equipment.<br>• Political risk insurance tailored for renewable energy and clean energy projects.<br>• Approval of more than $505 million to support up to six new global renewable energy investment funds and plans to support additional funds seeking to deploy equity capital in the sector.<br>• A new carbon policy to reduce greenhouse gas emissions from the OPIC project portfolio by 50 percent over a 15-year period. |
| U.S. Agency for International Development (USAID) | The U.S. Agency for International Development (USAID) is the principal U.S. agency that extends assistance to countries recovering from disaster, trying to escape poverty, and engaging in democratic reforms. USAID attempts to further U.S. foreign policy objectives by supporting economic growth, trade, agricultural development, global health, democracy, and conflict and humanitarian assistance. By recognizing the additional pressures climate change will place on low- and middle-income countries, expanding access to modern renewable energy and energy efficiency technologies is a cornerstone of the USAID energy mission. |
| U.S. Department of Commerce (DOC) | The U.S. Department of Commerce has a broad mandate to advance economic growth and to create jobs for the American people. The department has responsibilities in the areas of trade, technology, entrepreneurship, economic development, and statistical research and analysis.<br>DOC's International Trade Administration (ITA) is the lead trade-promotion agency of the U.S. government. ITA provides practical information to help businesses select markets and products, ensures that companies have access to international markets as required by U.S. trade agreements, and safeguards businesses from unfair competition because of dumped and subsidized imports. ITA houses four business units that help exporters. The U.S. and Foreign Commercial Service provides a point of contact for companies in 109 cities around the country and in more than 80 countries around the world to help U.S. companies enter and expand in foreign markets. The Manufacturing and Services unit provides industry specialists, including RE&EE specialists, that can identify trade opportunities for specific products or services. The Market Access and Compliance unit helps keep world markets open to U.S. products with a team of country specialists. The Import Administration identifies and pursues foreign unfair subsidies practices that impede U.S. exporters' access to foreign markets. |

| Agency | Description |
|---|---|
| | DOC is also home to the National Institute of Standards and Technology, which helps promote U.S. innovation and competitiveness by advancing measurement science, standards, and technology, and the National Oceanic and Atmospheric Administration, which can provide climate data to companies interested in particular markets. The Department of Commerce also chairs the Trade Promotion Coordinating Committee (TPCC) and co-chairs the TPCC Working Group on Renewable Energy and Energy Efficiency with the U.S. Department of Energy. |
| U.S. Department of Energy (DOE) | The U.S. Department of Energy seeks to advance the national, economic, and energy security of the United States; to develop technologies to mitigate the effects of climate change and to enable adaptation; to promote scientific and technological innovation in support of that mission; and to ensure the environmental cleanup of the national nuclear weapons complex. DOE works directly with foreign governments and institutions to promote dissemination of RE&EE technologies, with programs directed to international research and development, technical assistance and training, policy support, and market development assistance. DOE also co-chairs the TPCC Working Group on Renewable Energy and Energy Efficiency with the Department of Commerce. |
| U.S. Department of Labor | The U.S. Department of Labor does not have any programs specifically geared toward RE&EE exporting; however, the department's Employment and Training Administration (ETA) is aligning economic and workforce development programs to increase the availability of well-trained, highly skilled workers who can develop and produce green technologies and products. ETA has awarded more than $500 million in local and state grants as part of the American Recovery and Reinvestment Act of 2009. The grants focus on providing training for jobs in RE&EE, including the emerging electrical power, advanced drive-train vehicle, and biofuels industries. Although these "green jobs" grants were not designed specifically to increase exports, ETA supports small and medium-sized enterprises in training and hiring skilled workers, a linchpin to help keep their doors open and grow their businesses in the domestic and export sectors. |
| U.S. Department of State | The U.S. Department of State advances U.S. objectives and interests in the world by developing and implementing the president's foreign policy agenda. The Department of State is the lead U.S. agency in implementing the international components of policies to address climate change and to foster sustainable development. The Department of State undertakes a range of activities to promote the international deployment of RE&EE technologies, including the Energy and Climate Partnership of the Americas, the U.S.– China Energy Cooperation Program, and the Asia-Pacific Partnership on Clean Development and Climate, as well as the Major Economies Forum on Energy and Climate. |
| U.S. Department of Treasury | The U.S. Department of the Treasury is the lead cabinet agency for economic and financial issues. The department advises on domestic and international financial, monetary, economic, trade, and tax policy. The Department of the Treasury's Office of Environment and Energy was established in August 2008 to develop, coordinate, and execute the department's role in the domestic and international environment and energy agenda by addressing the broader economic issues related to climate and energy policies.<br>The Department of the Treasury supports the broader U.S. efforts to facilitate a |

| Agency | Description |
|---|---|
| | smooth transition to a national and global clean energy economy by leading U.S. environment and energy financing efforts in the G-20; by assisting the U.S. negotiating team on finance issues at the United Nations climate negotiations; and by managing U.S. interests at multilateral financial institutions, including the Climate Investment Funds and the Global Environmental Facility. |
| U.S. Environmental Protection Agency (EPA) | The U.S. Environmental Protection Agency (EPA) is responsible for implementing the environmental policy of the United States. The EPA is involved in a variety of international activities that support the deployment of RE&EE technologies, including a partnership with USAID that funds the U.S. Climate Technology Cooperative Gateway, a Web site with almost 1,000 resources on climate-friendly technologies (www.usctcgateway.gov). Additional EPA programs include technical support to the government of China to develop a voluntary energy efficiency–labeling program similar to Energy Star and the Wind Technology Partnership to accelerate the development and use of grid-connected wind power in China. |
| U.S. Small Business Administration (SBA) | The U.S. Small Business Administration (SBA) helps Americans start, build, and grow businesses in the United States and overseas. SBA offers a variety of financing options for small businesses, typically in the form of loan guarantees. SBA also provides counseling, training, business guidance, and support to small businesses for domestic and export activities. SBA offers three specialized loan guarantee programs to help provide export financing: the Export Express Program, the Export Working Capital Loan Program, and the International Trade Loan Program.<br>SBA maintains a network of Export Assistance Centers located in major metropolitan areas throughout the United States, which serve as one-stop shops for small and medium-sized enterprises in local markets.<br>Moreover, SBA maintains an export library, information on upcoming trade missions and events, and a trade links database that provides basic exporter information. |
| U.S. Trade and Development Agency | The U.S. Trade and Development Agency (USTDA) promotes economic development in low- and middle-income countries, as well as the export of U.S. products and services. USTDA funds various forms of technical assistance, early investment analysis, training, reverse trade missions, and business workshops that support the development of a modern infrastructure and a fair and open trading environment.<br>USTDA uses overseas grants, contracts with U.S. firms, and other funding tools to structure partnerships between U.S. companies and overseas project sponsors that bring proven U.S. technology, business practices, and expertise to the economic development of partner countries. In fiscal year 2009, USTDA prioritized activities promoting clean energy and climate change projects worldwide, committing almost $10 million in support to the sector—roughly 23.3 percent of total obligations. For fiscal year 2010, commitments increased to nearly $23 million, or nearly 50 percent of total obligations, and USTDA will maintain its emphasis on RE&EE projects in the years ahead. |
| U.S. Trade Representative | The U.S. Trade Representative (USTR) is a Cabinet member who serves as the President's principal trade advisor, negotiator, and spokesperson on trade issues. USTR coordinates trade policy, resolves disagreements, and frames issues for |

| Agency | Description |
|---|---|
| | presidential decision. USTR negotiates free trade agreements and Trade and Investment Framework Agreements that help to open foreign markets to U.S. exports. USTR's Office of Environment and Natural Resources has broad responsibilities to leverage trade negotiations and relationships to pursue environmental goals, including the reduction of trade barriers facing the RE&EE efficiency industries. |

## End Notes

[1] Executive Order No. 13534, 75 Federal Register 50 (March 11, 2010), pp. 12433-12435.

[2] U.S. export and import figures are calculated by the Department of Commerce based on a combination of data from DOC and information presented in U.S. Senator Ron Wyden, *Follow-up Report on Major Opportunities and Challenges to U.S. Exports to Environmental Goods,* May 2010.

[3] Federal Register Notice 103 (May 28,2010), pp. 29993-29994.

[4] *Bloomberg New Energy Finance.*

[5] United Nations Environment Programme Report, "Global Trends in Sustainable Energy Investment 2010: Analysis of Trends and Issues in the Financing of Renewable Energy and Energy Efficiency," Nairobi, Kenya, pp. 5.

[6] Database of State Incentives for Renewables and Efficiency, www.dsireusa.org

[7] The American Recovery and Reinvestment Act of 2009 allowed technologies usually qualifying for the production tax credit to choose the investment tax credit instead, with the option of substituting a cash grant instead of the investment tax credit. The cash grant option supports access to financing for companies struggling to find tax investors during the recession. In addition, more than $20 billion was made available for RE&EE technology investments through loan guarantees, grants, and research and development at the nation's energy research labs. The act's manufacturing tax credit provided a 30 percent tax credit for investments in 183 clean energy manufacturing facilities across 43 states. Although the purpose of the manufacturing tax credit was not to boost exports, the facilities could improve the export potential of U.S. RE&EE technologies, as well as the ability of U.S. companies to meet growing domestic demand

[8] This includes feasibility studies for model projects, basic and detailed design, manufacturing, installation, and operation of the project, and promotion of the technologies to wider communities in the country. New Energy and Industrial Technology Development Organization, "Outline of NEDO: New Energy and Industrial Technology Development Organization, 2008-2009," *www.nedo.go.jp/kankobutsu/pamphlets/kouhou/2008gaiyo_e/all.pdf,* pp. 138, 141, and 143.

[9] Japan, Spain, France, Denmark, the Netherlands, South Korea, Germany, Belgium, and Finland, listed in the order of the volume or reported official development assistance, starting with the largest.

[10] Based on information provided by Craig O'Connor, Export-Import Bank of the United States, June 18, 2010, with most information from OECD, "Review of the Sector Understanding on Export Credits, Renewable Energies, and Water Projects: July 2005 to June 2008," Paris, November 2008. Source of data for balance of 2008 and 2009 not identified.

[1]1 75 Federal Register Notice 103, June 22, 2010, pp. 35450-51.

[12] Department of Energy, "Renewable Energy Data Book 2009," pp. 82; Biomass Research and Development Technical Advisory Committee Meeting, National Renewable Energy Laboratory, 2006.

[13] U.S. export and import figures are calculated by the Department of Commerce based on a combination of data from the Department of Commerce and information presented to Senator Ron Wyden, *Follow-up Report on Major Opportunities and Challenges to U.S. Exports in Environmental Goods,* May 2010.

[14] American Wind Energy Association, "Small Wind Turbine Global Market Study," *www.awea.org/smallwind/pdf/2010_AWEA_Small_Wind_Turbine_Global_Market_Study.pdf*

[15] *Bloomberg New Energy Finance.*

[16] As of August 26, 2010, *Bloomberg New Energy Finance*

[17] Mark Taylor, "Bloomberg New Energy Finance," Research Note, "Geothermal Turbine Market Share," March 3, 2010, pp. 5.

[18] Renewable Energy Policy Network for the 21st Century, "Renewables 2010 Global Status Report," pp. 21.

[19] Energy Information Administration, "Geothermal Heat Pump Manufacturing Activities, 2008," December 2009, Figure 4.7.

[20] Renewable Energy Policy Network for the 21st Century, "Renewables 2010 Global Status Report," pp. 21.

[21] Ibid, pp. 21.

[22] The Energy Star Program does have a standard that measures energy efficiency in electronics, appliances, and some building materials, but this program is voluntary and does not represent all energy-efficient products.

[23] Senator Ron Wyden, "U.S. Trade of Environmental Goods: Follow-Up Report to Major Opportunities and Challenges to U.S. Exports of Environmental Goods," May 2010.

[24] For categories that included both RE&EE and non-RE&EE goods, the Wyden report estimated their respective shares in either value terms or as a percentage of the most specific level of trade product identified. The Wyden report then converted these estimates into percentage shares and values of U.S. exports and imports for the HTS 6-digit subheadings.

In: U.S. Renewable Energy and Energy ... ISBN: 978-1-61942-354-1
Editors: K. Norris and C. Steele 

*Chapter 2*

# RENEWABLE ENERGY AND ENERGY EFFICIENCY EXPORT INITIATIVE ON TRACK TO INCREASE RE&EE EXPORTS*

***Trade Promotion Coordinating Committee Working Group on Renewable Energy and Energy Efficiency***

## INTRODUCTION

The Trade Promotion Coordinating Committee Working Group on Renewable Energy and Energy Efficiency (RE&EE) is proud to announce that one year into the implementation of the Renewable Energy and Energy Efficiency Export Initiative (RE4I), every commitment has been achieved or is in the process of completion. Spurred by President Obama's challenge to position the United States as the leading exporter of clean energy technologies, the RE4I streamlined the process of obtaining financing for renewable energy exporters, dedicated additional resources to financing exports in the sector, and improved the delivery of existing export promotion

* This is an edited, reformatted and augmented version of a Trade Promotion Coordinating Committee Working Group on Renewable Energy and Energy Efficiency publication, dated October, 2011.

services for RE&EE companies – all in an era of scarce resources and budget uncertainty. In addition, the RE4I created a process for improving market access for RE&EE exporters and better targeting future trade promotion activities.

When President Obama announced the National Export Initiative (NEI) during his 2010 State of the Union Address, he did so in an effort to double total U.S. exports and support millions of jobs in all sectors and industries. Yet not all sectors have the same growth or export potential. High growth or high export potential sectors, like renewable energy and energy efficiency (RE&EE), health care, and biotechnology, have therefore been a focus of the NEI since its launch more than 18 months ago.

In his inaugural address, the President called for the expanded use of renewable energy to meet the twin challenges of energy security and climate change. The Recovery Act's investments in deployment of clean energy technologies and to enhance the RE&EE supply chain furthered this commitment. The Renewable Energy and Energy Efficiency Export Initiative (RE4I), a key component of the NEI, responded to the President's commitment by bringing together relevant federal agencies to improve the global competitiveness of U.S. RE&EE exporters.

This two-pronged approach – improving the U.S.- based RE&EE supply chain and enhancing the tools available to U.S. RE&EE exporters – ensures that U.S.

RE&EE companies are both ready and able to meet opportunities when they arise, either at home or abroad. The RE4I renewed the U.S. Government's commitment to eliminating trade barriers and enhancing export promotion; it strengthened the U.S. Government financing programs for U.S. RE&EE exporters; and it created a strong public/private dialogue to improve existing services and make future programs more effective.

This Annual Review details the progress of the RE4I since its inception in December 2010, providing success stories and an update on implementation of agency commitments. It is important to note that the commitments of the RE4I will continue through the NEI time frame and that, in addition to their implementation, the TPCC agencies will look for additional ways to promote exports in the sector.

# TAILORED FINANCING FOR THE SPECIFIC NEEDS OF THE RE&EE INDUSTRY

The first commitment under the RE4I is to better tailor export financing to meet the specific needs of the RE&EE industry. Without adequate financing, U.S. companies struggle to close deals, transport goods in a timely manner, or find foreign buyers. In fact, industry feedback received in response to a Federal Register solicitation for public comments often cited access to financing as the most significant barrier to increasing RE&EE exports.

Although the U.S. Government offered financing programs prior to the RE4I, the Initiative created several new programs focused specifically on facilitating private-sector investment in RE&EE technologies – both directly financing U.S. exports and driving demand for U.S. products and expertise.

## Streamlined Processing of RE&EE Financing Applications:

Both the U.S. Export-Import Bank (ExIm) and the Overseas Private Investment Corporation (OPIC) made a commitment in the RE4I to streamline their processing of RE&EE financing applications to shorten the amount of time it takes companies to access credit – increasing the ability of U.S. companies to close an export deal quickly.

Over the course of the year, ExIm reconfigured its Solar Express Program to include not only solar photovoltaic (PV), but "small wind" as well. The resulting Renewable Energy Express Program continues to generate tremendous interest, with applications for export finance coming from U.S. companies with projects across Asia, Latin America, the Caribbean and Europe. As an example, ExIm approved a $9 million loan guarantee in July 2011 to support the export of thin-f lm solar panels by Abound Solar to India, and a $22.2 million loan guarantee in May 2011 to support the export of small wind turbines by Northern Power Systems to Italy. [See Box 1]

Similarly, OPIC has also developed more streamlined processes for RE&EE financing applications under the auspices of the RE4I. OPIC now has a dedicated group for developing RE&EE investments, including the evaluation of debt financing and political risk insurance coverage. Examples of recent OPIC projects include a $121 million guaranty for the construction and operation of two 20 MW solar PV power plants in Peru; a $90 million loan for the expansion of biomass fuel production in Liberia; and a $58 million loan for the development, construction and operation of a 46.4 MW hydropower generation facility in Georgia.

## Greater OPIC and ExIm Focus on RE&EE Financing

Due to a more streamlined process for RE&EE applications and an organizational commitment to expanding investment in the sector, both ExIm and OPIC will invest more money in the RE&EE sector in 2011 than in any previous year. In 2011, ExIm will likely double its authorizations for RE&EE projects, supplying over $700 million worth of financing to foreign buyers of U.S. technologies. Similarly, OPIC has more than doubled its commitments to the RE&EE sector, climbing from $240 million in 2010 to likely commitments of $520 million in 2011, with projects ongoing in South America, Asia, and Africa. These increases in overall coverage have provided many RE&EE companies with a level playing field on which to compete.

## OPIC Support for the Leasing of U.S. Renewable Energy Equipment:

**Box 1. Export-Import Bank Creates Renewable Express Program**

A key RE4I commitment is to streamline the processing of renewable energy project applications at ExIm and OPIC. Responding to this commitment, ExIm's "Renewables Express" was created, expanding the successful Solar Express project finance review process to include additional renewable energy technologies. Under Renewables Express, ExIm will consider project financing for small renewable power producers seeking loans of $3 million to $10 million in an expedited review process for both corporate balance sheet and limited recourse transactions that fit within the program's parameters and where the repayment is generated from the project's cash flows.

OPIC is currently evaluating several potential transactions that would use its new leasing facility to enable U.S. companies to supply services for the construction and development of renewable energy projects. For example, U.S. geothermal firms that supply drilling and exploration services are very competitive globally. The ability to lease drilling rigs, rather than make the large investments needed to purchase them, will open new markets for U.S. geothermal exporters. The first of the transactions based on this leasing facility are expected to be completed by mid-2012.

**Box 2. OPIC's supplemental call for investment proposals will provide $500 million in OPIC coverage to mobilize more than $1.5 billion for RE&EE projects in emerging markets**

In December 2010, OPIC announced a Global Renewable Resources Funds call for proposals to help catalyze and facilitate private sector investments that promote renewable resources globally. OPIC ultimately approved five funds from among 56 respondents to the call. Additional funds are still under review.

In June 2011, OPIC's Board of Directors approved nearly $500 million in financing for five new investment funds that could foster more than $1.5 billion in private-sector investment in renewable energy projects in Southeast Asia and Africa. The new funds fulfilled a pledge made by OPIC President Elizabeth Littlefield at the UN Climate Change Conference in Cancun in December 2010, when she announced that OPIC would provide at least $300 million in financing for new private equity investment funds that could invest more than $1 billion in renewable resources projects in emerging markets.

For example, one of the new OPIC-supported investment funds – the Renewable Energy Asia Fund – will help deploy mature renewable energy technologies in India and the Philippines. The fund is aimed at decreasing the gap between energy supply and demand growth in these emerging economies while providing expertise to further the future development of renewable energy projects in the region.

## OPIC Commitment to Supplement Six Clean Energy Investment Funds with a New Call for Proposals:

OPIC's commitment to increasing private-sector investment in RE&EE projects was exemplified by its creation of new renewable energy investment funds. In June 2011, OPIC's Board of Directors approved nearly $500 million in financing for five new funds, which could support more than $1.5 billion in private-sector investment in renewable energy projects in Southeast Asia and Africa. [See Box 2]

## New OPIC Energy Efficiency Subordinated Debt Product:

Likewise, OPIC has developed a new subordinated debt product to support energy efficiency improvements and has begun marketing this opportunity to potential customers. The new subordinated debt product will finance 100 percent of a project's energy efficiency improvements and can be paid back with cost-savings over time. As an example, OPIC is currently providing $3 million in financing for a project in Jordan, which will accelerate the expansion of energy efficient services related to large scale solar water heating and cooling. OPIC will continue to shape the energy efficiency product based on feedback from industry.

## Increased USTDA Funding for Clean Energy Projects:

To fulfill a commitment made under RE4I, the U.S. Trade and Development Agency (USTDA) is continuing to prioritize funding for clean energy projects that promote U.S. exports. USTDA is on track to obligate approximately 40 percent of its total program for clean energy activities in Fiscal Year 2011. Examples of new clean energy projects funded by USTDA in the past year include feasibility studies on wind power and landfill gas projects in Mexico, a feasibility study on a waste-to-energy project in the Philippines, and two technical assistance projects in Morocco to support both photovoltaic and concentrated solar power development.

# Improved Market Access for U.S. RE&EE Exporters

The RE4I underscores the fact that U.S. firms routinely face at least one of two market access challenges when exporting. First, foreign governments have increasingly coupled strong policy incentives for renewable energy development with measures to shield their own manufacturers from foreign competition – thus limiting export opportunities either by barring imports from foreign countries, including the United States, or by incentivizing purchases from domestic suppliers. Second, many countries lack the foundational policy environment needed to encourage the deployment of RE&EE technologies. In these markets, despite renewable energy resource potential, the lack of policy forestalls the creation of a market for U.S. exports.

The RE4I's second commitment therefore intensifies efforts to create export markets for U.S. RE&EE companies. Whether by removing trade barriers or by fostering supportive policy environments, the RE4I brings a multi-agency, multi-faceted approach to creating demand for U.S. RE&EE products and services in foreign markets.

## TPCC-Led Renewable Energy Market Development Missions Organized by the Department of Commerce:

As an example of the RE4I's focus on generating opportunities for U.S. exporters, the Department of Commerce (DOC) pledged to lead market development missions to markets with significant resource potential for renewable energy, but which lack policy incentives to encourage project development. In September 2011, the Under Secretary of International Trade hosted a U.S.-Mexico Renewable Energy and Energy Efficiency Policy Roundtable in Mexico City. The event allowed U.S. companies to meet key decision-makers in the Mexican Government to discuss policy options to spur development and the potential to deploy U.S. renewable energy technologies in Mexico. USTDA, the Department of State (State), and the Department of Energy (DOE) assisted DOC in developing the Roundtable.

Market development missions are most effective when they target markets where significant adoption of renewable energy technologies is possible given a country's resource endowment of renewable energy potential. Mexico is a perfect example of this: the country enjoys world-class wind potential, significant solar radiation, and a wealth of geothermal, biomass, and hydropower potential. In addition, given Mexico's proximity to the United States, its status as a NAFTA partner, and the current volume of overall trade between the United States and Mexico, the DOC-led visit offers the potential to generate significant future demand for U.S. RE&EE exports. Over the course of the RE4I, DOC will lead similar market development missions to key markets, working with other TPCC agencies.

DOE also works to prime markets for new technologies. DOE's international activities help develop foreign markets for U.S. goods and services by providing expertise on policy development and quality control measures. As an example, in June 2011, DOE recognized a new international standard for energy management that will not only help consumers around the world save money, but also promote the deployment of energy efficiency technologies. The emergence of greater demand for these technologies is expected to provide an opportunity for U.S. exporters to reach new markets.

## Development of Standard RE&EE Technical Documentation to Expedite RE&EE Projects in Developing and Emerging Markets:

Because many emerging economies lack appropriate regulations and standard technical documentation to implement RE&EE projects, USTDA is providing technical assistance to help them develop these tools. For example, USTDA funded a standards development technical assistance project in China to help harmonize Chinese smart grid standards with U.S. smart grid equipment specifications and standards. Harmonized standards would help open China's smart grid market to U.S. companies – a potential market of $10 billion annually.

In addition to the smart grid standards project, USTDA funded a technical assistance project in Morocco that will determine the impact of a concentrated solar power plant on the electrical power grid, as well as evaluate power storage options and water usage. USTDA also funded two feasibility studies on developing tariff rate applications, power purchase agreements, and interconnection agreements that could be used to promote further development across Mexico's renewable energy market, providing increased opportunities for U.S. Firms.

### Creation of a Trade Policy Staff Committee (TPSC) Subcommittee Focused on RE&EE:

Earlier this year, the Office of the U.S. Trade Representative (USTR) established a new interagency TPSC Subcommittee on Environmental Goods and Services (Subcommittee). The Subcommittee, which met twice in 2011, is charged with identifying and addressing crosscutting trade and investment policy barriers that impact U.S. exporters of environmental goods and services. In addition to U.S. efforts in the World Trade Organization (WTO) and the Asia Pacific Economic Cooperation (APEC), the Subcommittee has focused its initial efforts on local content requirements that have been implemented or considered by several trading partners and their potential impact on U.S. exporters of environmental goods and services, in particular in the renewable energy sector.

## ENHANCED INFORMATION AND TRADE PROMOTION EFFORTS TO LINK BUYERS AND SELLERS OF U.S. RE&EE PRODUCTS AND SERVICES

Once markets demand RE&EE products and services, U.S. companies in these sectors must establish relationships with foreign customers. The third commitment of the RE4I is to link buyers and sellers of U.S. RE&EE technology. By showcasing U.S. products and services, introducing U.S. companies to export opportunities, and focusing on those markets most likely to result in U.S. exports, the RE4I ensures that U.S. RE&EE exporters have the fact-based, targeted trade promotion services they need to be successful.

### Increased Number of RE&EE Trade Missions:

Trade missions are often the most effective way of getting U.S. RE&EE exporters connected with foreign buyers. Despite budget limitations, DOC has pursued an aggressive series of programs to support RE&EE exports and has increased the number of RE&EE trade missions over 2010 levels. Nineteen DOC missions in 2011 included a renewable energy or energy efficiency component; most of these involved some level of TPCC coordination. An upcoming trade mission to Turkey is a perfect example. In December 2011,

DOC will lead several U.S. renewable energy and energy efficiency companies to Izmir, Turkey to build on the considerable accomplishments of the "Near-Zero" project, which involved coordination between DOE, DOC, State, USTDA, ExIm, and OPIC. [See Box 3]

**Box 3. Energy Efficiency Export Opportunities Increased Through "Near Zero Zone" Project in Izmir, Turkey**

In 2011, the Department of Energy-led Near-Zero Zone project in Izmir, Turkey made substantial progress in mobilizing energy efficiency export opportunities. The Near-Zero Zone project is designed to demonstrate the positive impact lower energy consumption can have on business productivity, while also promoting U.S. energy efficiency exports.

Following a baseline energy audit, a minimum of three U.S. energy efficiency equipment and service exporters conducted energy efficiency technical assessments on 20 Turkish manufacturers operating in the Izmir Ataturk Organized Industrial Zone. USTDA provided funding for 10 of the technical assessments. The Turkish companies will have the opportunity to purchase technologies to realize the energy savings identified through the energy efficiency technical assessments.

In December, the Department of Commerce will lead an energy efficiency trade mission to Turkey, including the Izmir industrial zone, helping to link U.S. exporters to the Turkish companies interested in buying technology from the United States. Ex-Im and OPIC are working to identify facilities to finance such purchases.

## Additional RE&EE Reverse Trade Missions and Pre-Feasibility Studies:

USTDA's reverse trade missions bring foreign project sponsors who are interested in buying RE&EE technologies to the United States, exposing foreign buyers to U.S. exporters. Continuing its commitment to the RE4I, USTDA sponsored 16 clean energy reverse trade missions in FY 2011, including a solar power reverse trade mission from southern Africa. Several U.S. companies expect business to develop from this mission, with at least one estimating future sales of U.S.-manufactured equipment valued at $50 million.

Similarly, USTDA sponsored a delegation of Brazilian smart grid officials to visit the United States; six U.S. companies reported that the visit benefitted their commercial relationships in Brazil, with one company already projecting sales of U.S.-manufactured equipment and services of up to $40 million. [See Box 4]

**Box 4. U.S. Trade and Development Agency Reverse Trade Missions Spur Exports**

Bringing buyers to the United States to meet with U.S. RE&EE technology vendors and to see the technology in operation is an important component of USTDA's efforts under the RE4I strategy. In this endeavor, USTDA has supported and continues to fund multiple renewable energy reverse trade missions in solar, wind, geothermal, waste-to-energy and hydropower.

One outcome of such efforts is nearly $50 million in total U.S. exports in 2011 for a wind power project in Brazil that resulted from a Latin American and Caribbean regional wind energy reverse trade mission in 2010. Although the sales to the wind power project had been under consideration before the visit, the Brazilian officials cemented their commitment to proceed with the transaction based on the trade mission.

## Expansion of the Green Embassies Program:

As a way of demonstrating U.S. RE&EE technologies in foreign markets, DOC and State developed the League of Green Embassies, a global network of diplomatic missions that have committed to showcase U.S. RE&EE technologies used in renovation and new-build projects in embassies and consulates around the world. By using the products and services to meet the embassies' own energy reduction targets, participating U.S. missions introduce host-country officials and foreign buyers to innovative U.S. products. Though the RE4I only called for the Green Embassies Program to expand to 12 embassies, today the program includes over 70 embassies and consulates worldwide.

The program has moved beyond the demonstration stage and has already resulted in several "beyond the-fence" exports in LED technology, micro-wind, and window ~ lm products. WindStream Technologies, for example, closed a $10 million deal in Brazil for 30,000 small wind turbines in part as a result of the Green Embassies Program. In 2012, ten additional embassy makeovers will be conducted in Europe, with additional renovations planned for Latin America. [See Box 5]

**Box 5. The Departments of Commerce and State Expand the Green Embassies Program**

Though the RE4I committed the Departments of Commerce and State to expand the Green Embassies Program to 12 embassies or consulates, the program has experienced tremendous growth in 2011 and now exceeds 70 diplomatic missions around the world. The program provides U.S. RE&EE companies with a one-of-a-kind opportunity to demonstrate their technology in foreign markets, by installing equipment in U.S. embassies. The embassy is then able to showcase available U.S. technology to foreign buyers and foreign government officials.

The first U.S. embassy to undergo a Green Embassies makeover was the Ambassador's residence in Brussels, Belgium. The project included upgrading indoor and outdoor lighting, installing small wind turbines, and improving the energy efficiency of windows and doorways. Eight U.S. companies contributed technology to the makeover, which provided an immediate energy savings of 24 percent. Importantly, several of the U.S. companies that demonstrated their technology have already reported export sales as a result of the Green Embassies Program. As the program expands further and additional makeovers are accomplished, future exports will also likely increase as a result of the program.

## DOC Study Rating Top Renewable Energy Markets for Potential U.S. Exports:

In an era of scarce government resources, it is critical that the U.S. Government focus export promotion services on the most promising markets for U.S. exporters. DOC is developing a "best market prospects" study to help U.S. Government agencies target trade promotion programs on short, medium, and long term export opportunities by country and technology within the renewable energy sector. Using trade and energy data, a qualitative assessment of existing policy environments, and feedback from a survey of commercial and economic officers in U.S. embassies around the world, DOC is developing a methodology for identifying trade targets and better directing the RE4I's implementation.

The results of the "best market prospects" study will be available to other U.S. Government agencies to help plan trade missions, trade events, reverse trade missions, and determine market priorities by the end of 2011.

## Foreign Buyers' Guides to U.S. Renewable Energy Products and Services:

Foreign buyers need access to information on U.S. renewable energy technologies and services. Some foreign governments, like Japan and Australia, support web-based information that enables potential customers to identify relevant vendors and products for purchase. The United States o ers no comparable source of centralized information.

Under the RE4I, DOE assessed the potential for U.S. renewable energy associations to provide specific company or product information by linking their web-based information to export.gov, the U.S. Government's export portal. Certain associations representing RE&EE exporters supported linking association information to export.gov, but recommendations on implementation differed based on the size of the association and the interests of member companies. Going forward, the DOC will examine the potential to implement the associations' recommendations, taking into account existing resources.

## DOE Assessment of the Export Potential of Products and Services Related to Energy Efficiency:

Similarly, the United States has no centralized source of information on energy efficiency products and services available for export and little information on which energy efficiency products and services are viable in global markets. DOE is addressing this need by performing an analysis that segments global markets according to critical drivers of demand for energy efficient products and services, such as climate zone, manufacturing pro les, energy costs and sources, and commitment to GHG reductions. This analysis will help identify sub-markets and corresponding definitions, a critical step in assessing the potential for U.S. export opportunities. The study is scheduled to be completed in 2012.

## Boost U.S. RE&EE Exports through the EcoPartnerships Program:

To further demonstrate U.S. clean technology in key markets, State, along with several TPCC partner agencies, continues to use the EcoPartnerships Program to increase U.S. competitiveness in the fast-growing China market. The EcoPartnerships Program provides a platform to promote solutions to China's environmental challenges, creating new markets for the U.S. clean technology industry.

Under the RE4I, State aimed to sponsor five to ten new collaborations annually. In May of 2011, six new EcoPartnerships were signed that created opportunities in seven U.S. states by linking U.S. companies with Chinese partners in the development of cutting-edge technology. For instance, the State of Utah and Qinghai Province announced the creation of a new joint research center in Utah. In addition to helping develop technology in the United States, the partnership created a market in China for the technology that is eventually developed at the research center.

## Expansion of U.S. Department of Agriculture's (USDA) Foreign Agricultural Service's (FAS) Market Access Program to Include Wood Pellets:

U.S. exports of wood pellets are expanding rapidly. In 2010, the United States exported $128 million (736,000 MT) of wood pellets, a 37 percent increase over 2009. To further facilitate this export growth, the Department of Agriculture's FAS welcomed its first biomass energy cooperator into the Market Access Program in 2011. The American Biomass Trade Cooperative (ABTc) received its FAS Market Access Program (MAP) startup funding January 1, 2011. The award is being used to conduct market research in the European Union to facilitate U.S. wood pellet exports to the region.

In addition to partnering with the new biomass cooperator, FAS's Global Policy Analysis Division expanded its Biofuels Gain market research report with a section on "Biomass for Heat and Power" to include production, consumption, and trade statistics for wood pellets.

## STRENGTHEN U.S. GOVERNMENT SERVICES FOR U.S. RE&EE COMPANIES

The RE4I is a true interagency initiative and therefore successful implementation is dependent on coordination across agencies, as well as continued dialogue with U.S. RE&EE exporters. Information continues to be shared across departments and export opportunities jointly pursued between policy-makers in Washington, DC and at embassies, missions, and export assistance centers around the world.

During the development of the RE4I, several companies underscored the need for greater dialogue with the U.S. Government, and highlighted their difficulty identifying suitable trade promotion programs as a barrier to export competitiveness. As a result, the RE4I's final commitment focuses on ensuring that the U.S. Government solicits industry advice regularly and provides U.S. exporters with improved information on existing trade promotion programs. This effort is ongoing, but significant progress has been made already.

### Creation of a RE&EE Advisory Committee to Provide Private-Sector Expertise to the Secretary of Commerce:

To ensure that the U.S. Government receives up-to-date advice and expertise from the RE&EE industry, the RE4I created a Renewable Energy and Energy Efficiency Advisory Committee. Consisting of 27 senior-level officials from across the RE&EE industry, the Committee met for the first time on December 7, 2010 and was tasked by then-Secretary of Commerce Gary Locke to develop specific, actionable recommendations for the Administration to improve export competitiveness in the sector.

The Committee met again in March, June, August and September 2011. The Committee announced 11 initial recommendations on September 15, 2011. The recommendations focused on establishing a sustainable U.S. clean energy market, expanding potential exporters' access to finance, and addressing gaps in information on global RE&EE markets and U.S. competitiveness. The RE&EEAC will continue to evaluate and make recommendations through the end of its current charter in July 2012.

## Develop a Better Methodology for Tracking U.S. RE&EE Exports:

It is difficult to evaluate the RE4I's effect on the global competitiveness of the U.S. RE&EE industry and to make specific recommendations on additional policies and programs to boost exports due to the absence of reliable export figures. For this reason, one important aspect of the RE4I was the establishment of a baseline for judging the RE4I's success.

DOC, working with the International Trade Commission and DOE, is in the process of developing a more robust methodology for tracking RE&EE exports and benchmarking U.S. export competitiveness in key export markets. Analysts at DOC and DOE have worked to develop a more comprehensive list of products used to produce renewable energy. The resulting schematic will likely clarify the renewable energy portion of existing Harmonized Tariff System codes and provide a fuller picture of exports in the sector.

## Single Web Portal for RE&EE Exporters:

During the development of the RE4I, several companies requested a single, easy-to-use guide to existing export promotion programs. In response, one of the RE4I's first achievements was the debut of a TPCC-wide web portal that provides information from across the U.S. Government on RE&EE market research, news, and upcoming events hosted by export.gov, which is managed by DOC.

The U.S. government often employed social networking to expand public awareness of the RE4I in 2011. An April re-tweet of the original RE4I report by the White House on Twitter highlighted the initiative to over two million people. The RE4I was featured on the White House blog, the Department of Commerce blog, and websites from several U.S. Government agencies. The result was an increase in exposure for the RE4I and the dissemination of information to a greater number of U.S. companies.

The portal has drawn considerable interest from across the RE&EE industry; in total, 16,300 visitors have used the web portal since its launch, 18 percent of whom were repeat visitors. Additionally, the total number of page views on the portal has exceeded 95,300. Starting in June 2011, DOC began supplementing the portal with a monthly RE&EE Exporters' Update, which provides alerts on new information on the web portal to more than 1,500 email subscribers.

## Online Exporters' Guide for U.S. RE&EE Companies:

The RE4I also provides U.S. companies with a comprehensive online guide to existing U.S. Government export promotion services. The guide offers answers to frequently asked questions on several agencies' programs and provides links to specific financing, trade facilitation, and export promotion websites from across the U.S. Government. It also describes examples of U.S. RE&EE companies that have used the programs successfully. The online guide will reflect any future changes to federal export promotion programs, providing U.S. exporters with updated information.

## U.S. Government RE&EE Exporters Brochure:

To highlight the web portal and online guide, DOC, with support from DOE and input from TPCC agencies, also developed a printed brochure that is now distributed widely at RE&EE trade shows and conferences. The brochure includes a matrix matching agencies' programs with typical services sought by U.S. exporters and directs companies to the interagency portal.

## Unified U.S. Government Outreach at Key RE&EE Trade Shows:

To demonstrate the interagency nature of U.S. Government RE&EE trade promotion, the TPCC Working Group on RE&EE moved quickly to provide U.S. companies with a one-stop-shop for information on U.S. Government export financing, trade promotion, and policy development through coordinated outreach at key industry trade shows. The TPCC agencies had already been cooperating extensively through the Department of Commerce's International Buyers Program (IBP) events. The IBP program brings foreign buyers to the United States to participate in U.S. trade shows. Through this program, small and medium-sized enterprises can effectively meet foreign buyers without traveling overseas. As the IBP program expands to include additional RE&EE shows, such as Windpower and Green Build, opportunities for further coordination will increase.

**Box 6. A Recap of RE4I Deliverables**

Eight TPCC agencies committed to 23 new actions or programs to help meet the goals of the RE4I. Each deliverable has been implemented within existing budgets and within existing authorities. Following is a recap of the RE4I's deliverables is below:

Tailored Financing for RE&EE Companies:

- Streamlined agency processing of RE&EE transactions at both Ex-Im and OPIC
- Greater OPIC and ExIm focus on RE&EE financing
- OPIC support for U.S. leasing sector of RE&EE equipment
- OPIC commitment to supplement the $505 million in financing for six clean energy investment funds announced in 2008
- New energy efficiency subordinated debt product from OPIC
- Increased USTDA funding for clean energy projects

Improved Market Access for U.S. RE&EE Companies

- TPCC-led renewable energy and energy efficiency market development missions organized by the Department of Commerce
- Development of standard RE&EE technical documentation through USTDA to expedite RE&EE projects in developing and emerging markets
- Creation of a TPSC subcommittee focused on RE&EE by the Office of the U.S. Trade Representative

Enhanced Information and Trade Promotion Efforts to Link Buyers and Sellers of U.S. RE&EE Products and Services

- Increased number of RE&EE trade missions
- Additional USTDA-funded RE&EE reverse trade missions and pre-feasibility studies
- Expansion of the Green Embassies Program
- Department of Commerce "top prospects" study on RE&EE export markets
- Foreign buyers guides to U.S. RE&EE technology developed by the Departments of Commerce and Energy
- Department of Energy study on the export potential of energy efficiency products and services
- Boosting RE&EE exports through the EcoPartnerships Program
- Expansion of USDA's Foreign Agricultural Service's Market Development Program to include wood pellets

Strengthen U.S. Government Services for U.S. RE&EE Companies

- Creation of a RE&EE Advisory Committee to provide private-sector expertise to the Secretary of Commerce
- Develop a better methodology for tracking U.S. RE&EE exports
- Single web portal for RE&EE exporters
- Online exporters guide for U.S RE&EE companies
- Development of a U.S. Government RE&EE exporters brochure
- Unified U.S. Government outreach at key RE&EE trade shows

In: U.S. Renewable Energy and Energy ... ISBN: 978-1-61942-354-1
Editors: K. Norris and C. Steele 

*Chapter 3*

# U.S. TRADE IN ENVIRONMENTAL GOODS FOLLOW-UP REPORT TO MAJOR OPPORTUNITIES AND CHALLENGES TO U.S. EXPORTS OF ENVIRONMENTAL GOODS*

***Ron Wyden***

May 20, 2010

## SUMMARY

The United States and its trading partners have put a high priority on domestic and international efforts to reduce emissions of greenhouse gases, mitigate climate change, and protect the environment. In the face of these efforts, those countries that develop and foster competitive environmental goods industries will be in the best position to reap the economic rewards of increased international demand for goods that mitigate climate-related and other environmental challenges.

In December 2009, my office produced a special report entitled Major Opportunities and Challenges to U.S. Exports of Environmental Goods. The report showed that (1) global trade of environmental goods

* This is an edited, reformatted and augmented version of the Unite.States Trade in Environmental Goods Follow-up Report given by Senator Ron Wyden to Major Opportunities and Challenges to the United States Exports of Environmental Goods, on May 20, 2010.

more than doubled in recent years; (2) although U.S. exports of environmental goods increased, the U.S. trade balance of these goods deteriorated substantially; (3) trade barriers in the fastest growing markets abroad are a key constraint to increasing U.S. exports; and (4) U.S. exporters are losing overseas market share to foreign producers of environmental goods, especially to China.

A key constraint of that report and all others that preceded it was the lack of reliable trade data of goods that serve a specific environmental purpose. The World Bank has identified 43 products as broadly being "climate friendly" (the WB 43); however, accurate trade data for these products are largely unavailable because these goods are commingled with other non-environmental goods classified within the same broader product categories that are recognized at the international level. The WB 43 is generally viewed as a basis for international trade negotiations to reduce or eliminate trade barriers to environmental goods. Until today, there was not a clear understanding of the share of these 43 goods that serve a specific (or "real") environmental purpose in relation to the broader product groupings in which these goods are classified at the international level.

This report presents a new approach to refine data related to U.S. trade in environmental goods to give policy makers a clearer understanding about the opportunities and challenges to American producers and exporters in this sector of the economy. Building on Major Opportunities and Challenges to U.S. Exports of Environmental Goods, this report estimates U.S. trade flows of real environmental goods within the broader product groupings in which these goods are classified.

Key highlights of this report show:

- U.S. trade levels in these environmental goods are overstated. U.S. exports of environmental goods are much smaller than previously estimated, accounting for only 7–12 percent ($1.3–1.9 billion) of previous estimates of U.S. exports between 2007 and 2009.
- The U.S. trade deficit of these environmental goods is 30 percent smaller than previously thought. Products associated with renewable energy, such as wind and solar power, account for 80 percent of the U.S. trade deficit.
- Despite running an overall negative trade balance in environmental goods, the United States runs a positive trade balance in goods associated with wastewater treatment, air pollution control, and heat and energy management.
- Many states produce and export environmental goods and services. Future efforts to reduce or remove foreign trade barriers and promote U.S. exports of environmental goods and services would therefore provide broad-based benefits to U.S. industry across the country.

# Background

Environmental goods, along with environmental services (collectively referred to as "EGS"), cover a wide range of products and services that cut across many different industry sectors. Although there is no internationally accepted definition, EGS are generally defined as goods and services associated with environmental protection, including those related to air, water, or soil pollution control and prevention; waste management; environmental monitoring and recycling; and renewable energy, among others.

EGS were singled out in Paragraph 31 (iii) of the World Trade Organization's (WTO) Doha negotiating mandate.[1] While the mandate calls for a reduction, or as appropriate, elimination of tariffs and non-tariff barriers, progress in the negotiations is hampered by the absence of universally accepted definitions of EGS and the lack of a common understanding of the scope of goods and services to be liberalized under that mandate.

Members of the WTO continue efforts to reach consensus on what constitutes an environmental good (henceforth "EG"). In December 2007, the United States and the EU proposed an Environmental Goods and Services Agreement (EGSA) to liberalize trade in environmentally friendly goods and services.[2] The proposed plurilateral agreement called for elimination of tariffs and identified non-tariff barriers of 43 products identified by the World Bank as broadly being "climate friendly." The proposed EGSA would also seek commitments to establish binding tariff elimination commitments on an additional, broader set of products that were not specifically identified in the proposal, as well as enhanced market access for trade in environmental services.

A central issue in these negotiations is whether expanded trade liberalization should include only those goods with a single use (e.g., environmental protection or climate change mitigation), or should also include dual-use goods that have both environmental and non-environmental uses. This issue is complicated by the fact that both single- and dual-use goods are often included in broader WTO-recognized 6- digit code levels of the Harmonized Commodity Description and Coding System (HS), which include products generally considered not to be EG.

These conceptual and practical issues have hindered the development of data and estimates of trade flows of EG. As a result, a clear understanding of the potential effect on trade flows of any liberalization of trade is lacking. Previous work has focused on the trade flows of the broader HS 6-digit subheadings where global trade data are more readily available.

This report estimates U.S. trade flows of the more narrowly defined product list of 43 environmental goods contained within the broader HS 6-digit subheadings. For the purposes of this report, the term "real environmental goods" is used to distinguish the more narrowly defined environmental goods from all goods contained within the broader HS 6-digit subheadings.

Environmental goods are categorized into the following 7 product groups:

1) Air Pollution Control (APC)
2) Management of Solid and Hazardous Waste (SHW)
3) Renewable Energy Products (REP)
4) Heat and Energy Management (HEM)
5) Waste Water Management and Potable Water Treatment (WWM)
6) Cleaner or More Resource Efficient Technologies and Products (CRET)
7) Environmental Monitoring, Analysis, and Assessment Equipment (EMAA)

## METHODOLOGY

This report estimates the U.S. trade balance, U.S. exports, and U.S. imports of the more narrowly defined product list of 43 "real" environmental goods contained within the broader HS 6-digit subheadings. In contrast to the previous report, this report does not estimate the global trade of the more narrowly defined list of environmental goods, nor does it estimate the U.S. market share of the more narrowly defined list of environmental goods in global markets.

Trade data for the latest 3-year period (2007-09) of the 43 HS 6-digit subheadings were obtained from the U.S. International Trade Commission.[3]

Where the product list of 43 environmental goods corresponded to a specific HS 6-digit or 10-digit level in the U.S. Harmonized Tariff Schedule (HTS) or U.S. Export Schedule B (SB), official trade data were provided for that product. In the majority of cases, trade was estimated in either value terms or as a percentage of the most specific level of trade nomenclature identified. These estimates were then converted to percentage shares and values of U.S. exports and imports for the HS 6-digit subheadings.

Estimates of the value of U.S. trade and the percentage of the HS 6-digit subheadings are presented in a detailed summary table in Annex A. An aggregated summary table is presented in Annex B. Product descriptions, states engaged in the production and export of environmental goods, principal U.S. export markets, and principal foreign producers and exporters are presented in Annex C.

Estimates of U.S. trade of the 43 environmental goods were made by review of publications and industry data, and through consultations with government officials, industrial officials, and members of nongovernmental organizations and industry trade associations. Substantial assistance was provided by staff of the U.S. International Trade Commission.

## U.S. MERCHANDISE TRADE BALANCE

Between 2007 and 2009, the U.S. trade balance in EG deteriorated, but was tempered by reduced EG imports in 2009 as a result of the U.S. recession and reduced U.S. demand. The negative U.S. trade balance was approximately $4 billion in 2009, accounting for about 70 percent of previous estimates of the negative U.S. trade balance based on HS 6-digit subheadings of these products (figure 1).

Renewable energy products account for a significant share of the negative U.S. trade balance (figure 2). U.S. trade in wind towers, wind turbines, and solar cells and modules accounted for 80 percent of the U.S. trade deficit of environmental goods in 2009 (figure 3). Excluding renewable energy products, the U.S. trade balance was positive for products associated with water waste management (WWM), air pollution control (APC), and heat and energy management (HEM) (figure 4).

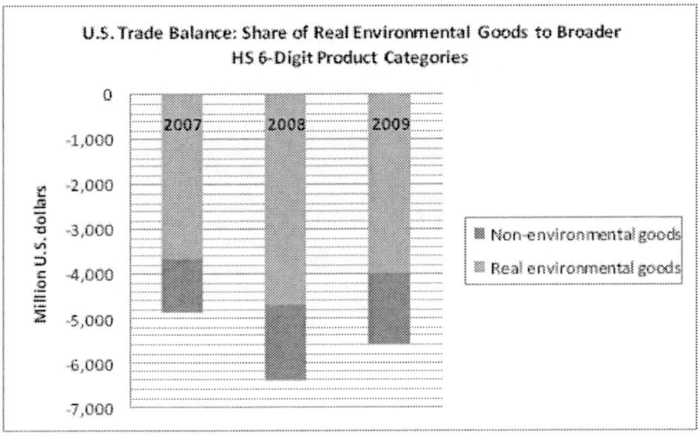


Figure 1.

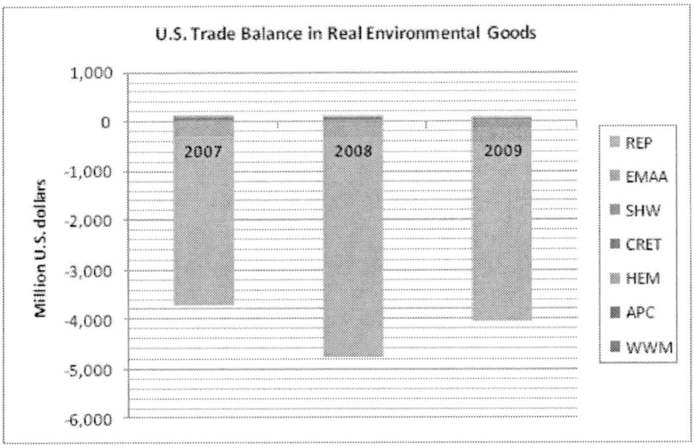


Product Key: APC: Air Pollution Control; SHW: Management of Solid and Hazardous Waste;REP: Renewable Energy Products; HEM: Heat and Energy Management; WWM: Waste Water Management and Treatment; CRET: Cleaner or More Resource Efficient Technologies; EMAA: Environmental Monitoring Equipment.

Figure 2.

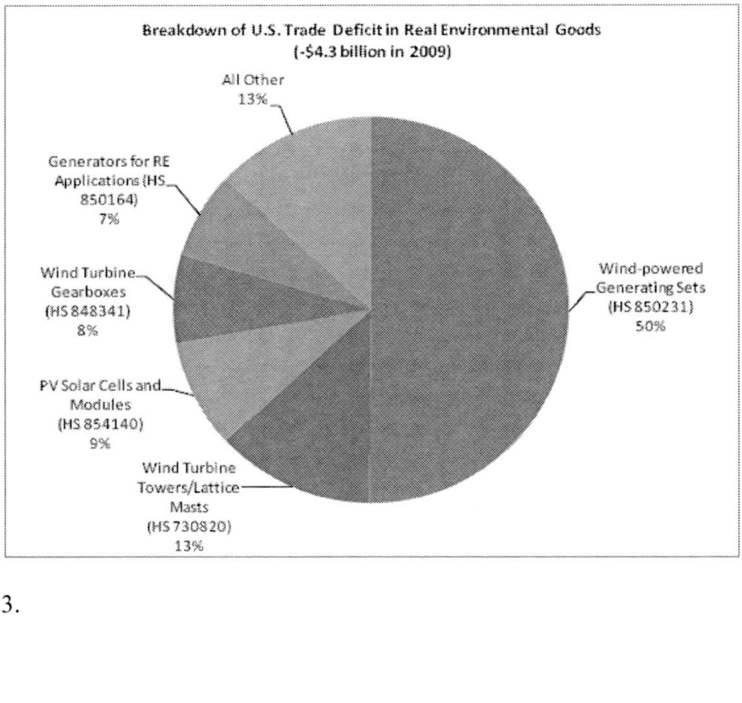


Figure 3.

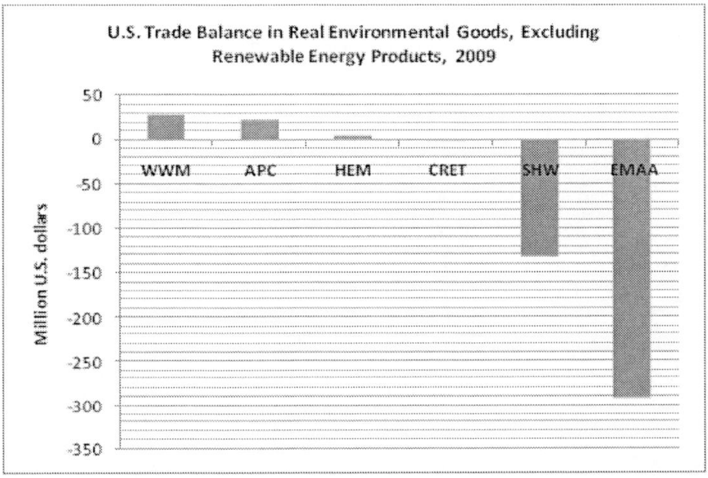


Product Key: APC: Air Pollution Control; SHW: Management of Solid and Hazardous Waste REP: Renewable Energy Products; HEM: Heat and Energy Management; WWM: Waste Water Management and Treatment, CRET: Cleaner or More Resource Efficient Technologies; EMAA: Environmental Monitoring Equipment.

Figure 4.

## U.S. EXPORTS

Between 2007 and 2009, estimated U.S. EG exports accounted for 7–12 percent of U.S. exports of all products in the 43 HS 6-digit subheadings (figure 5). Percentage shares vary by product group (figure 6). In some cases, U.S. exports of a specific environmental good are substantially smaller than previously thought. For example, gas turbines for use in clean energy systems are estimated to account for only 0.1 percent ($3.4 million in 2009) of U.S. exports of all the products classified within HS 8411.82, which was previously used to estimate exports of this particular environmental good (see appendix A). During the 2007–09 period, estimated U.S. EG exports increased almost 50 percent to $1.9 billion (figure 7). Renewable energy products such as solar cells and modules, wind turbines, solar water heaters, and generators for renewable energy applications accounted for 70 percent of estimated U.S. EG exports in 2009 (figure 8).

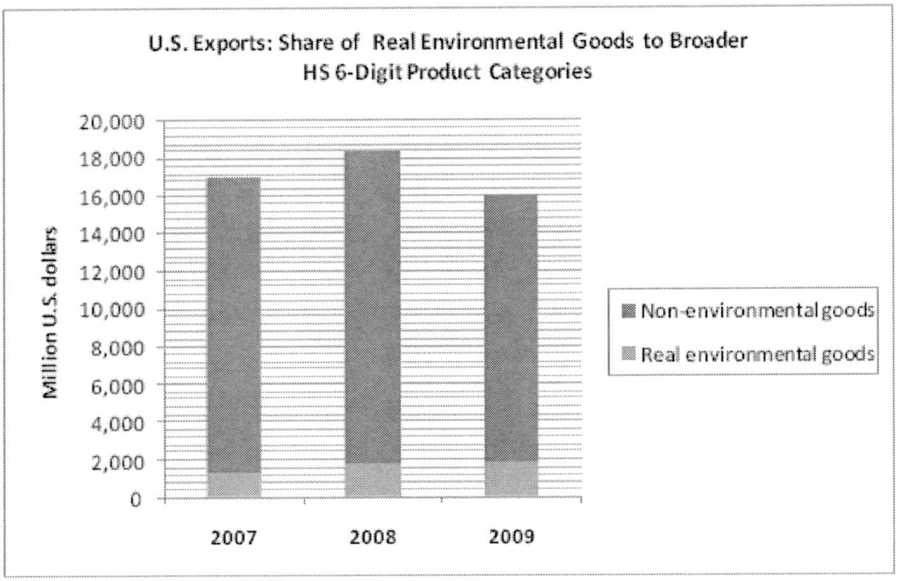


Figure 5.

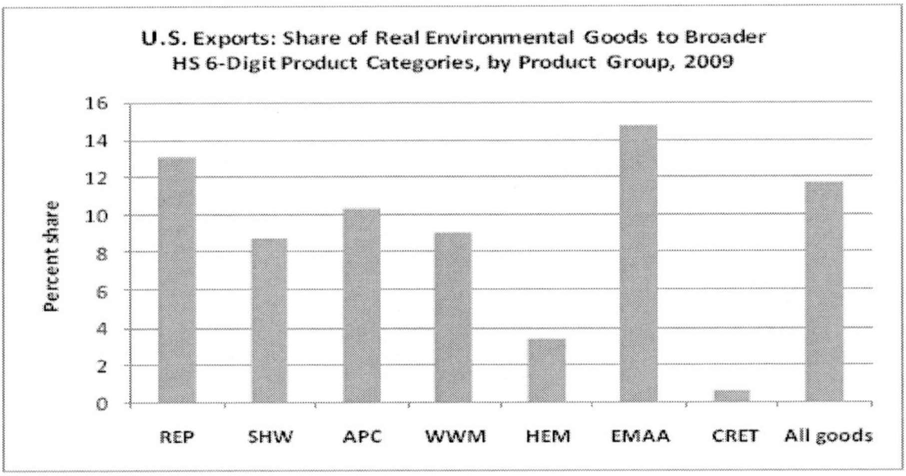


Product Key: APC: Air Pollution Control; SHW: Management of Solid and Hazardous Waste; REP: Renewable Energy Products; HEM: Heat and Energy Management; WWM: Waste Water Management and Treatment; CRET: Cleaner or More Resource Efficient Technologies; EMAA: Environmental Monitoring Equipment.

Figure 6.

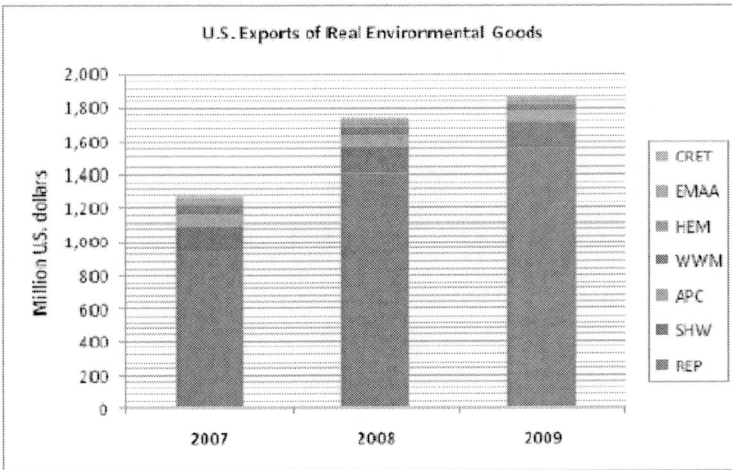


Product Key: APC: Air Pollution Control; SHW: Management of Solid and Hazardous Waste; REP: Renewable Energy Products, HEM: Heat and Energy Management, WWM: Waste Water Management and Treatment; CRET: Cleaner or More Resource Efficient Technologies; EMAA: Environmental Monitoring Equipment.

Figure 7.

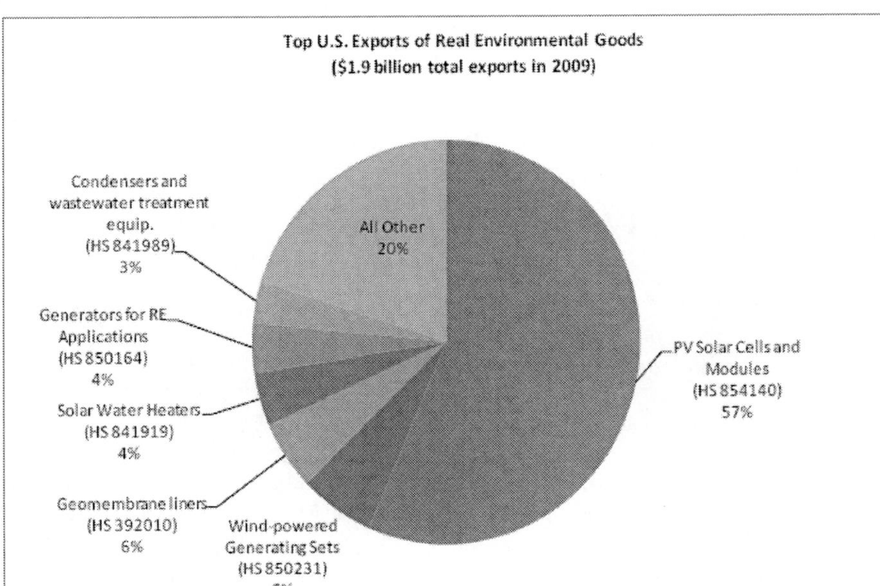


Figure 8.

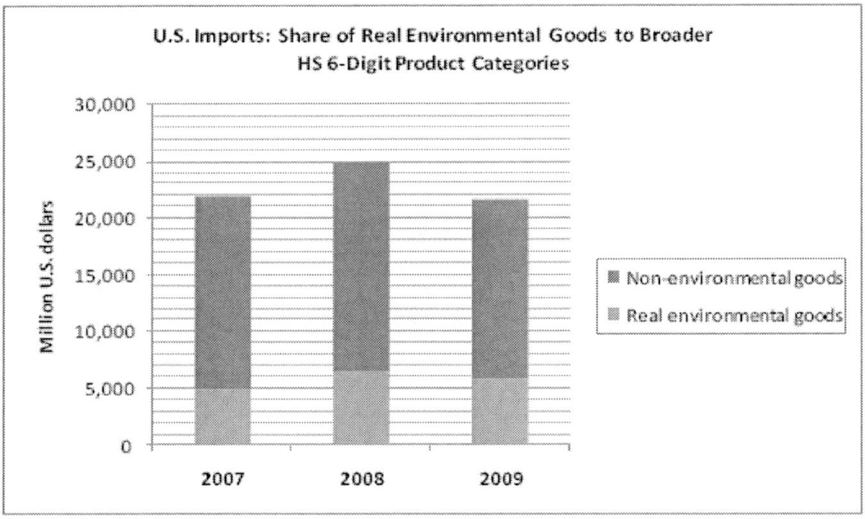


Figure 8'.

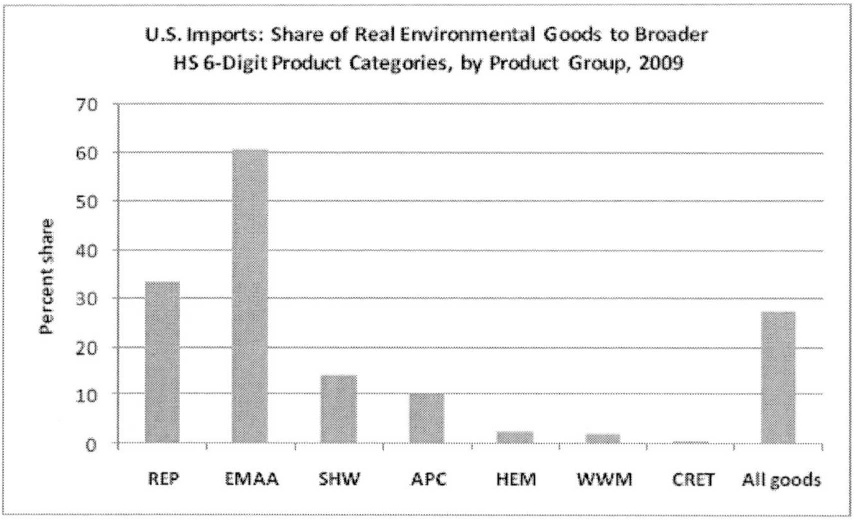


Product Key: APC: Air Pollution Control; SHW: Management of Solid and Hazardous Waste; REP: Renewable Energy Products; HEM: Heat and Energy Management; WWM: Waste Water Management and Treatment; CRET: Cleaner or More Resource Efficient Technologies; EMAA: Environmental Monitoring Equipment.

Figure 9.

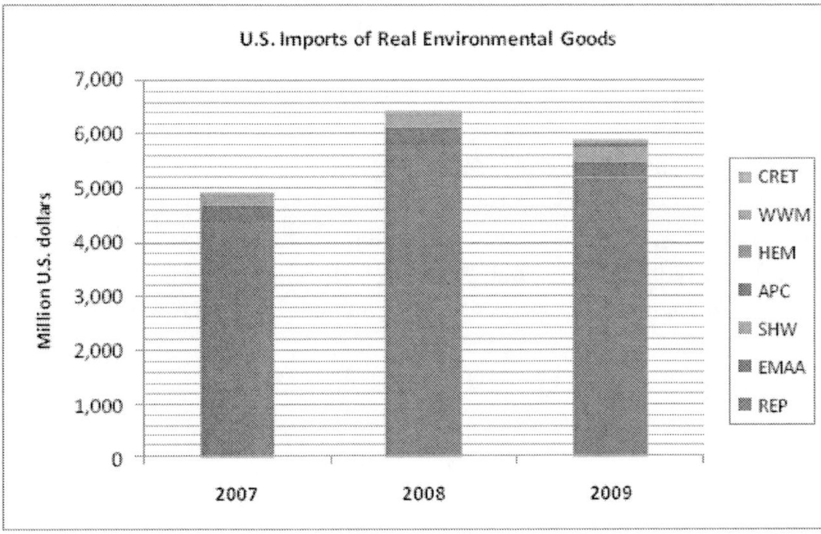


Product Key: APC: Air Pollution Control SHW: Management of Solid and Hazardous Waste; REP: Renewable Energy Products; HEM: Heat and Energy Management; WWM: Waste Water Management and Treatment; CRET: Cleaner or More Resource Efficient Technologies; EMAA: Environmental Monitoring Equipment.

Figure 10.

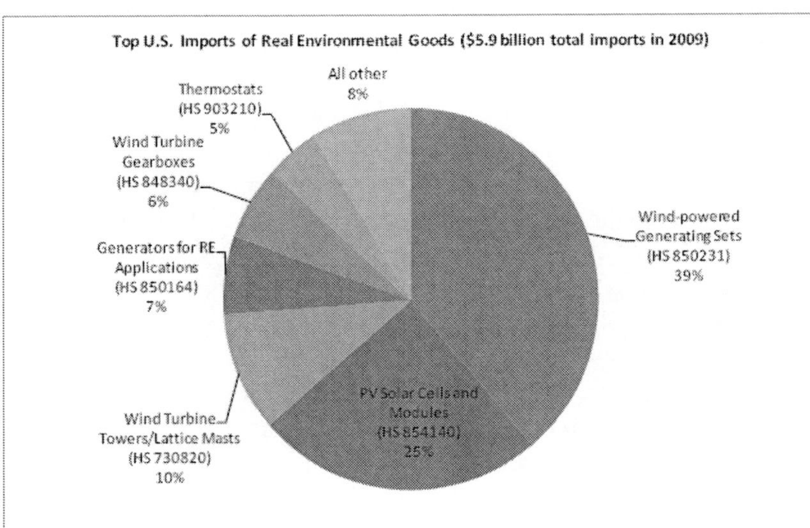


Figure 11.

## U.S. IMPORTS

Between 2007 and 2009, estimated U.S. EG imports accounted for 23–27 percent of U.S. imports of all products in the 43 HS 6-digit subheadings (figure 8'). Percentage shares vary by product group (figure 9). During the 2007–09 period, estimated U.S. EG imports increased almost 20 percent to $5.9 billion (figure 10). Renewable energy products such as wind turbines, wind towers, and generators for renewable energy applications accounted for almost 90 percent of estimated U.S. EG imports in 2009 (figure 11).

## CONCLUSION

This report shows that U.S. exports of environmental goods are substantially smaller than previously thought. Such a large discrepancy between the new estimates of U.S. trade data presented in this report and previous estimates suggests than an accurate understanding of U.S. trade flows of environmental goods is lacking. As the Congress considers extension of investment tax credits for manufacturing facilities associated with renewable energy, the way it incentivizes the deployment of environmental goods, and the President's request to fund the National Export Initiative, it must also consider how to accurately measure and assess U.S. trade in environmental goods.

Because this report has shown the considerable variation in what may or may not constitute an environmental good in U.S. trade data classified and aggregated at a level that is comparable at an international level, Congress should consider giving the United States Trade Representative (USTR) more explicit guidance about the scope of a proposed EGSA in order to ensure that the Congress's expectations are consistent with the Administration's, since any EGSA will need congressional ratification.

This report is the first effort to offer a new approach to measuring U.S. trade in environmental goods. It separates out those goods with a specific or *real* environmental purpose from non-environmental goods that are classified in the same broader product groupings. The approach used in this report could serve the basis for further efforts to refine trade data of environmental goods in order to address the current U.S. trade imbalance, assess U.S. export competitiveness, and further engage WTO members to agree on a specific definition of environmental goods. A specific definition and a standardized approach to measure and delineate environmental goods from non-environmental goods that are classified within the same product groupings could help clarify negotiating positions in any potential multilateral agreement to liberalize trade in environmental goods.

U.S. agencies responsible for trade matters should improve collaboration in their efforts to analyze U.S. trade in environmental goods. The approach used in this report could serve as a basis for further efforts to refine trade data. To this end, the Office of the USTR should continue its recent work with the U.S. International Trade Commission and the U.S. Department of Commerce to develop a better definition of and better data on U.S. trade of environmental goods. This will assist the government's ability to determine the potential economic effects on the U.S. economy of liberalizing trade in environmental goods through any potential plurilateral agreement, such as the U.S.-proposed EGSA.

Until the United States and its trading partners are better able to determine how to categorize and account for environmental goods within their respective tariff schedules, U.S.-led efforts to liberalize trade in an agreed-upon set of environmental goods will prove difficult, given the nature of the international classification system. In addition, although not explicitly touched upon in this report, efforts to define, classify, and measure trade in environmental services will prove equally challenging. Discussions about this report may bring to light these challenges. Despite them, the United States should continue to play an active leadership role at the international level to foster cooperation in this important area of trade.

# ANNEX A.

## Estimates of U.S. trade of the 43 environmental goods, 2007-09

| HS 6-digit code | Label | 2007 | | | 2008 | | | 2009 | | |
|---|---|---|---|---|---|---|---|---|---|---|
| | | Product trade ($millions) | HS 6-digit trade ($millions) | Share (%) | Product trade ($millions) | HS 6-digit trade ($millions) | Share (%) | Product trade ($millions) | HS 6-digit trade ($millions) | Share (%) |
| Renewable Energy Plant (REP) | | | | | | | | | | |
| 730820 | Imports | 555.6 | 578.8 | 96 | 925.5 | 944.4 | 98 | 599.8 | 612.2 | 98 |
| | Exports | 51.0 | 85.1 | 60 | 41.9 | 91.1 | 45 | 39.3 | 65.5 | 60 |
| | Balance | -504.6 | -493.7 | – | -883.4 | -853.4 | – | -560.5 | -546.7 | – |
| 761100 | Imports | 0.0 | 4.1 | 0 | 0.0 | 2.4 | – | 0.0 | 2.3 | 0 |
| | Exports | 0.0 | 18.8 | 0 | 0.0 | 22.7 | – | 0.0 | 13.8 | 0 |
| | Balance | 0.0 | 14.7 | – | 0.0 | 20.3 | – | 0.0 | 11.4 | – |
| 840681 | Imports | 3.5 | 57.6 | 6 | 0.4 | 39.5 | 1 | 1.9 | 31.9 | 6 |
| | Exports | 0.0 | 3.2 | 0.1 | 0.0 | 13.1 | 0.1 | 0.1 | 78.2 | 0.1 |
| | Balance | -3.5 | -54.4 | – | -0.4 | -26.4 | – | -1.8 | 46.3 | – |
| 841011 | Imports | 0.5 | 0.5 | 100 | 1.9 | 1.9 | 100 | 5.5 | 5.5 | 100 |
| | Exports | 3.8 | 3.8 | 100 | 6.7 | 6.7 | 100 | 3.8 | 3.8 | 100 |
| | Balance | 3.3 | 3.3 | – | 4.8 | 4.8 | – | -1.7 | -1.7 | – |
| 841090 | Imports | 0.2 | 36.3 | 1 | 1.0 | 45.2 | 2 | 2.7 | 49.1 | 6 |
| | Exports | 1.9 | 28.6 | 7 | 3.4 | 35.1 | 10 | 1.9 | 50.5 | 4 |
| | Balance | 1.7 | -7.6 | – | 2.4 | -10.2 | – | -0.8 | 1.4 | – |
| 841181 | Imports | 0.1 | 148.5 | 0.1 | 0.2 | 213.9 | 0.1 | 0.3 | 330.7 | 0.1 |
| | Exports | 0.6 | 592.5 | 0.1 | 0.7 | 738.5 | 0.1 | 0.4 | 390.4 | 0.1 |
| | Balance | 0.4 | 444.0 | – | 0.5 | 524.6 | – | 0.1 | 59.7 | – |
| 841182 | Imports | 0.2 | 171.1 | 0.1 | 0.2 | 231.5 | 0.1 | 0.2 | 195.6 | 0.1 |
| | Exports | 3.4 | 3,381.0 | 0.1 | 3.1 | 3,140.6 | 0.1 | 2.9 | 2,834.0 | 0.1 |
| | Balance | 3.2 | 3,209.9 | – | 2.9 | 2,909.2 | – | 2.7 | 2,638.4 | – |
| 841581 | Imports | 1.0 | 24.4 | 4 | 1.0 | 32.4 | 3 | 1.0 | 52.9 | 2 |
| | Exports | 5.0 | 109.3 | 5 | 8.0 | 143.0 | 6 | 10.0 | 140.8 | 7 |
| | Balance | 4.0 | 84.9 | – | 7.0 | 110.5 | – | 9.0 | 87.8 | – |

| HS 6-digit code | Label | 2007 | | | 2008 | | | 2009 | | |
|---|---|---|---|---|---|---|---|---|---|---|
| | | Product trade | HS 6-digit trade | Share | Product trade | HS 6-digit trade | Share | Product trade | HS 6-digit trade | Share |
| | | ($millions) | ($millions) | (%) | ($millions) | ($millions) | (%) | ($millions) | ($millions) | (%) |
| **Renewable Energy Plant (REP)** | | | | | | | | | | |
| 841861 | Imports | 10.0 | 15.9 | 63 | 10.0 | 51.1 | 20 | 10.0 | 56.6 | 18 |
| | Exports | 35.0 | 77.9 | 45 | 47.0 | 76.5 | 61 | 60.0 | 80.3 | 75 |
| | Balance | 25.0 | 62.0 | – | 37.0 | 25.4 | – | 50.0 | 23.7 | – |
| 841869 | Imports | 0.0 | 812.2 | 0 | 0.0 | 918.4 | 10 | 0.0 | 831.2 | 0 |
| | Exports | 0.0 | 849.5 | 0 | 0.0 | 928.0 | 9.1 | 0.0 | 705.1 | 0 |
| | Balance | 0.0 | 37.3 | – | 0.0 | 9.7 | – | 0.0 | -126.2 | – |
| 841919 | Imports | 14.0 | 360.2 | 3.9 | 30.7 | 383.4 | 8 | 11.2 | 328.1 | 3.4 |
| | Exports | 23.1 | 92.3 | 25 | 71.1 | 114.7 | 62 | 78.7 | 119.3 | 66 |
| | Balance | 9.0 | -267.9 | – | 40.4 | -268.7 | – | 67.6 | -208.8 | – |
| 841990 | Imports | 5.8 | 720.0 | 0.8 | 7.5 | 829.2 | 0.9 | 7.8 | 652.3 | 1.2 |
| | Exports | 3.5 | 588.3 | 0.6 | 3.3 | 655.8 | 0.5 | 4.5 | 564.2 | 0.8 |
| | Balance | -2.2 | -131.9 | – | -4.2 | -173.3 | – | -3.3 | -88.1 | – |
| 848340 | Imports | 278.0 | 1,859.7 | 15 | 431.3 | 2,219.3 | 19 | 365.5 | 1,619.4 | 23 |
| | Exports | 25.1 | 743.0 | 3 | 28.3 | 885.9 | 3 | 44.1 | 732.8 | 6 |
| | Balance | -252.9 | -1,116.7 | – | -402.9 | -1,333.4 | – | -321.5 | -886.6 | – |
| 848360 | Imports | 16.1 | 295.4 | 5 | 28.8 | 305.5 | 9 | 22.4 | 198.9 | 11 |
| | Exports | 2.2 | 215.7 | 1 | 2.4 | 235.4 | 1 | 2.1 | 203.9 | 1 |
| | Balance | -13.9 | -79.6 | – | -26.5 | -70.1 | – | -20.4 | 5.0 | – |
| 850161 | Imports | 13.9 | 138.9 | 10 | 17.0 | 169.7 | 10 | 11.5 | 114.7 | 10 |
| | Exports | 14.0 | 70.0 | 20 | 18.9 | 94.4 | 20 | 15.8 | 79.1 | 20 |
| | Balance | 0.1 | -68.9 | – | 1.9 | -75.3 | – | 4.3 | -35.6 | – |
| 850162 | Imports | 5.2 | 52.2 | 10 | 7.1 | 70.6 | 10 | 5.2 | 51.9 | 10 |
| | Exports | 22.1 | 110.5 | 20 | 23.7 | 118.3 | 20 | 24.3 | 121.3 | 20 |
| | Balance | 16.9 | 58.3 | – | 16.6 | 47.6 | – | 19.1 | 69.4 | – |

## Annex A. (Continued)

| HS 6-digit code | Label | 2007 | | | 2008 | | | 2009 | | |
|---|---|---|---|---|---|---|---|---|---|---|
| | | Product trade | HS 6-digit trade | Share | Product trade | HS 6-digit trade | Share | Product trade | HS 6-digit trade | Share |
| | | ($millions) | ($millions) | (%) | ($millions) | ($millions) | (%) | ($millions) | ($millions) | (%) |
| **Renewable Energy Plant (REP)** | | | | | | | | | | |
| 850163 | Imports | 7.0 | 69.6 | 10 | 8.1 | 81.4 | 20 | 3.7 | 36.7 | 20 |
| | Exports | 2.2 | 10.8 | 20 | 2.1 | 10.4 | 20 | 2.0 | 10.0 | 20 |
| | Balance | -4.8 | -58.9 | – | -6.1 | -71.0 | – | -1.7 | -26.7 | – |
| 850164 | Imports | 392.8 | 561.2 | 70 | 577.7 | 825.4 | 70 | 385.9 | 551.3 | 70 |
| | Exports | 60.4 | 302.1 | 20 | 83.9 | 419.1 | 20 | 74.1 | 370.5 | 20 |
| | Balance | -332.1 | -259.1 | – | -493.9 | -406.3 | – | -311.8 | -180.9 | – |
| 850231 | Imports | 2,379.9 | 2,379.9 | 100 | 2,503.3 | 2,503.3 | 100 | 2,280.0 | 2,280.0 | 100 |
| | Exports | 14.2 | 14.2 | 100 | 22.1 | 22.1 | 100 | 117.0 | 117.0 | 100 |
| | Balance | -2,365.8 | -2,365.8 | – | -2,481.3 | -2,481.3 | – | -2,163.0 | -2,163.0 | – |
| 850720 | Imports | 10.0 | 550.3 | 1 | 10.0 | 642.2 | 2 | 10.0 | 481.8 | 1 |
| | Exports | 20.0 | 388.3 | 5 | 20.0 | 520.9 | 4 | 20.0 | 374.8 | 5 |
| | Balance | 10.0 | -162.0 | – | 10.0 | -121.3 | – | 10.0 | -107.0 | – |
| 853710 | Imports | 4.8 | 4,829.2 | 0.1 | 5.1 | 5,082.9 | 0.1 | 4.0 | 4,030.9 | 0.1 |
| | Exports | 2.3 | 2,303.3 | 0.1 | 2.3 | 2,341.2 | 0.1 | 1.9 | 1,911.1 | 0.1 |
| | Balance | -2.5 | -2,526.0 | – | -2.7 | -2,741.7 | – | -2.1 | -2,119.8 | – |
| 854140 | Imports | 676.9 | 2,111.9 | 32 | 1,240.8 | 2,727.4 | 45 | 1,445.7 | 2,584.6 | 56 |
| | Exports | 646.9 | 1,582.2 | 41 | 1,011.3 | 1,976.1 | 51 | 1,056.7 | 2,017.6 | 52 |
| | Balance | -30.0 | -529.7 | – | -229.5 | -751.3 | – | -389.0 | -567.0 | – |
| 900190 | Imports | 0.3 | 332.1 | 0.1 | 0.4 | 365.6 | 0.1 | 0.3 | 275.6 | 0.1 |
| | Exports | 1.3 | 1,344.3 | 0.1 | 0.9 | 874.8 | 0.1 | 0.9 | 848.0 | 0.1 |
| | Balance | 1.0 | 1,012.2 | – | 0.5 | 509.2 | – | 0.6 | 572.4 | – |
| 900290 | Imports | 0.1 | 114.1 | 0.1 | 0.1 | 137.4 | 0.1 | 1.3 | 129.2 | 0.1 |
| | Exports | 0.1 | 99.1 | 0.1 | 0.1 | 86.9 | 0.1 | 0.6 | 62.8 | 0.1 |
| | Balance | 0.0 | -15.0 | – | 0.0 | -50.5 | – | -0.7 | -66.4 | – |
| **Subtotal** | Imports | 4,375.9 | 16,224.1 | 27 | 5,808.1 | 18,824.0 | 31 | 5,175.9 | 15,503.4 | 33 |
| | Exports | 938.1 | 13,013.8 | 7 | 1,401.2 | 13,551.3 | 10 | 1,561.1 | 11,894.8 | 13 |
| | Balance | -3,437.8 | -3,210.3 | – | -4,406.9 | -5,272.7 | – | -3,614.8 | -3,608.6 | – |

| HS 6-digit code | Label | 2007 | | | 2008 | | | 2009 | | |
|---|---|---|---|---|---|---|---|---|---|---|
| | | Product trade ($millions) | HS 6-digit trade ($millions) | Share (%) | Product trade ($millions) | HS 6-digit trade ($millions) | Share (%) | Product trade ($millions) | HS 6-digit trade ($millions) | Share (%) |
| **Environmental Monitoring, Analysis, and Assessment Equipment (EMAA)** | | | | | | | | | | |
| 903210 | Imports | 277.6 | 500.2 | 55 | 291.7 | 526.9 | 55 | 275.7 | 475.7 | 58 |
| | Exports | 5.5 | 109.3 | 5 | 5.6 | 112.0 | 5 | 4.5 | 90.3 | 5 |
| | Balance | -272.1 | -390.9 | – | -286.1 | -414.9 | – | -271.2 | -385.3 | – |
| 903220 | Imports | 39.7 | 39.7 | 100 | 45.6 | 45.6 | 100 | 31.5 | 31.5 | 100 |
| | Exports | 19.9 | 19.9 | 100 | 14.4 | 14.4 | 100 | 10.3 | 10.3 | 100 |
| | Balance | -19.8 | -19.8 | – | -31.1 | -31.1 | – | -21.2 | -21.2 | – |
| Subtotal | Imports | 317.3 | 539.9 | 59 | 337.3 | 572.5 | 59 | 307.2 | 507.2 | 61 |
| | Exports | 25.4 | 129.2 | 20 | 20.0 | 126.4 | 16 | 14.8 | 100.6 | 15 |
| | Balance | -291.9 | -410.7 | – | -317.3 | -446.1 | – | -292.4 | -406.6 | – |
| **Management of Solic and Hazardous Waste (SHW)** | | | | | | | | | | |
| 392010 | Imports | 59.3 | 921.3 | 6.4 | 57.9 | 992.0 | 5.8 | 42.9 | 872.8 | 4.9 |
| | Exports | 123.8 | 955.3 | 13 | 127.1 | 1,189.5 | 10.7 | 103.4 | 990.0 | 10.4 |
| | Balance | 64.5 | 34.0 | – | 69.2 | 197.5 | – | 60.5 | 117.2 | – |
| 761290 | Imports | 0.0 | 194.4 | 0 | 0.0 | 203.4 | 0 | 0.0 | 175.7 | 0 |
| | Exports | 1.5 | 347.5 | 0.4 | 1.5 | 333.8 | 0.4 | 1.5 | 371.3 | 0.4 |
| | Balance | 1.5 | 153.1 | – | 1.3 | 130.5 | – | 1.5 | 195.5 | – |
| 840219 | Imports | 1.0 | 10.3 | 10 | 1.8 | 18.3 | 10 | 7.5 | 74.6 | 10 |
| | Exports | 3.0 | 29.7 | 10 | 3.6 | 36.3 | 10 | 2.7 | 27.2 | 10 |
| | Balance | 1.9 | 19.4 | – | 1.8 | 18.0 | – | -4.7 | -47.4 | – |
| 840290 | Imports | 27.4 | 274.1 | 10 | 29.8 | 298.3 | 10 | 37.4 | 374.4 | 10 |
| | Exports | 17.0 | 170.4 | 10 | 22.1 | 220.8 | 10 | 23.7 | 237.1 | 10 |
| | Balance | -10.4 | -103.7 | – | -7.7 | -77.5 | – | -13.7 | -137.3 | – |
| 840410 | Imports | 4.8 | 48.1 | 10 | 11.0 | 110.4 | 10 | 10.9 | 109.0 | 10 |
| | Exports | 2.8 | 27.6 | 10 | 2.9 | 29.4 | 10 | 2.8 | 28.2 | 10 |
| | Balance | -2.0 | -20.4 | – | -8.1 | -81.0 | – | -8.1 | -80.8 | – |
| 841940 | Imports | 64.3 | 142.8 | 45 | 105.1 | 233.5 | 45 | 186.4 | 414.3 | 45 |
| | Exports | 11.0 | 52.3 | 21 | 16.1 | 89.7 | 18 | 18.5 | 92.4 | 20 |
| | Balance | -53.3 | -90.5 | – | -88.9 | -143.7 | – | -168.0 | -321.9 | – |
| **Subtotal** | Imports | 156.8 | 1,591.0 | 10 | 205.6 | 1,855.9 | 11 | 285.1 | 2,020.8 | 14 |
| | Exports | 159.1 | 1,582.8 | 10 | 173.3 | 1,899.5 | 9 | 152.6 | 1,746.2 | 9 |
| | Balance | 2.3 | -8.2 | – | -32.3 | 43.6 | – | -132.5 | -274.6 | – |

## Annex A. (Continued)

| HS 6-digit code | Label | 2007 | | | 2008 | | | 2009 | | |
|---|---|---|---|---|---|---|---|---|---|---|
| | | Product trade ($millions) | HS 6-digit trade ($millions) | Share (%) | Product trade ($millions) | HS 6-digit trade ($millions) | Share (%) | Product trade ($millions) | HS 6-digit trade ($millions) | Share (%) |
| Air Pollution Control (APC) | | | | | | | | | | |
| 840490 | Imports | 11.7 | 116.7 | 10 | 12.9 | 128.6 | 10 | 15.3 | 152.9 | 10 |
| | Exports | 5.5 | 54.6 | 10 | 4.6 | 46.5 | 10 | 5.9 | 59.1 | 10 |
| | Balance | -6.2 | -62.1 | – | -8.2 | -82.1 | – | -9.4 | -93.8 | – |
| 840510 | Imports | 0.0 | 24.4 | 0 | 0.0 | 15.9 | 0 | 31.2 | 37.0 | 84 |
| | Exports | 0.0 | 58.3 | 0 | 0.0 | 65.4 | 0 | 0.0 | 78.8 | 0 |
| | Balance | 0.0 | 33.8 | – | 0.0 | 49.5 | – | -31.2 | 41.8 | – |
| 841989 | Imports | 0.5 | 346.1 | 0 | 0.5 | 373.5 | 0 | 0.5 | 269.6 | 0 |
| | Exports | 62.5 | 518.2 | 11 | 62.5 | 836.4 | 8 | 62.5 | 525.9 | 11 |
| | Balance | 62.0 | 172.1 | – | 62.0 | 462.9 | – | 62.0 | 256.3 | – |
| **Subtotal** | Imports | 12.2 | 487.2 | 3 | 13.4 | 518.0 | 3 | 47.0 | 459.5 | 10 |
| | Exports | 68.0 | 631.1 | 11 | 67.1 | 948.3 | 7 | 68.4 | 663.8 | 10 |
| | Balance | 55.8 | 143.9 | – | 53.7 | 430.3 | – | 21.4 | 204.3 | – |
| Heat and Energy Management (HEM) | | | | | | | | | | |
| 701931 | Imports | 9.0 | 73.6 | 12 | 7.0 | 56.9 | 12 | 4.0 | 43.9 | 9 |
| | Exports | 12.0 | 84.2 | 14 | 14.0 | 117.5 | 12 | 12.0 | 100.0 | 12 |
| | Balance | 3.0 | 10.6 | – | 7.0 | 60.6 | – | 8.0 | 56.1 | – |
| 841950 | Imports | 9.5 | 477.4 | 2 | 14.5 | 722.7 | 2 | 16.2 | 809.0 | 2 |
| | Exports | 13.4 | 668.2 | 2 | 15.3 | 764.9 | 2 | 13.0 | 651.5 | 2 |
| | Balance | 3.8 | 190.8 | – | 0.8 | 42.2 | – | -3.2 | -157.5 | – |
| **Subtotal** | Imports | 18.5 | 551.0 | 3 | 21.5 | 779.6 | 3 | 20.2 | 852.9 | 2 |
| | Exports | 25.4 | 752.4 | 3 | 29.3 | 882.4 | 3 | 25.0 | 751.5 | 3 |
| | Balance | 6.9 | 201.4 | – | 7.8 | 102.8 | – | 4.8 | -101.4 | – |
| Waste Water Management and Potable Water Treatment (WWM) | | | | | | | | | | |
| 560314 | Imports | 1.0 | 58.0 | 1.7 | 1.0 | 57.3 | 1.7 | 1.0 | 82.5 | 1.2 |
| | Exports | 1.0 | 165.2 | 0.6 | 1.0 | 206.5 | 0.5 | 1.0 | 156.7 | 0.6 |
| | Balance | 0.0 | 107.2 | – | 0.0 | 149.2 | – | 0.0 | 74.2 | – |
| 730900 | Imports | 12.5 | 227.1 | 6 | 11.3 | 308.9 | 4 | 12.5 | 587.3 | 2 |
| | Exports | 42.5 | 253.9 | 17 | 38.3 | 379.3 | 10 | 42.5 | 268.4 | 16 |
| | Balance | 30.0 | 26.8 | – | 27.0 | 70.4 | – | 30.0 | -318.9 | – |
| 732490 | Imports | 3.6 | 177.8 | 2 | 3.9 | 194.5 | 2 | 2.9 | 149.2 | 2 |
| | Exports | 0.7 | 66.7 | 1 | 0.7 | 73.6 | 1 | 0.6 | 65.0 | 1 |
| | Balance | -2.9 | -111.1 | – | -3.2 | -120.9 | – | -2.3 | -84.2 | – |
| **Subtotal** | Imports | 17.1 | 462.9 | 4 | 16.2 | 560.7 | 3 | 16.4 | 819.0 | 2 |
| | Exports | 44.2 | 485.8 | 9 | 40.0 | 659.4 | 6 | 44.1 | 490.1 | 9 |
| | Balance | 27.1 | 22.9 | – | 23.8 | 98.7 | – | 27.7 | -328.9 | – |

| Product Key | | 2007 | | | 2008 | | | 2009 | | |
|---|---|---|---|---|---|---|---|---|---|---|
| | | EG product trade | HS 6-digit trade | Share | EG product trade | HS 6-digit trade | Share | EG product trade | HS 6-digit trade | Share |
| | | ($ millions) | ($ millions) | (%) | ($ millions) | ($ millions) | (%) | ($ millions) | ($ millions) | (%) |
| WWM | Imports | 17.1 | 462.9 | 4 | 16.2 | 560.7 | 3 | 16.4 | 819.0 | 2 |
| | Exports | 44.2 | 485.8 | 9 | 40.0 | 659.4 | 6 | 44.1 | 490.1 | 9 |
| | Balance | 27.1 | 22.9 | | 23.8 | 98.7 | | 27.7 | -328.9 | |
| CRET | Imports | 4.0 | 1,987.5 | 0.2 | 3.9 | 1,714.2 | 0.2 | 2.9 | 1,385.0 | 0.2 |
| | Exports | 2.9 | 384.1 | 0.8 | 2.4 | 360.6 | 0.7 | 1.9 | 322.9 | 0.6 |
| | Balance | -1.1 | -1,603.4 | | -1.5 | -1,353.6 | | -1.0 | -1,062.1 | |
| | | | | | | | | | | |
| Totals | Imports | 4,901.8 | 21,843.6 | 22 | 6,406.0 | 24,824.9 | 26 | 5,854.7 | 21,547.8 | 27 |
| | Exports | 1,263.1 | 16,979.2 | 7 | 1,733.3 | 18,427.9 | 9 | 1,867.9 | 15,969.9 | 12 |
| | Balance | -3,638.7 | -4,864.4 | | -4,672.7 | -6,397.0 | | -3,986.8 | -5,577.9 | |

Note.--The HS 6-digit trade values are official U.S. trade data; the values of imports and exports and the shares of the HS 6-digit trade are upper bound estimates of the trade of the specific products. Numbers may not add due to rounding.

Product Key:

APC: Air Pollution Control
SHW: Management of Solid and Hazardous Waste
REP: Renewable Energy Products
HEM: Heat and Energy Management
WWM: Waste Water Management and Treatment
CRET: Cleaner or More Resource Efficient Technologies
EMAA: Environmental Monitoring Equipment

# ANNEX C

## Environmental goods production description, state production, U.S. export markets, and foreign producers and exporters

| HS 6-digit code | Product description | States where there is production and/or assembly | Principal U.S. export markets | Principal foreign producers and exporters |
|---|---|---|---|---|
| Renewable Energy Plant (REP) | | | | |
| 730820 | Towers and lattice masts for wind turbines | Illinois, Iowa, North Dakota, Oklahoma, Texas | Jordan, Spain, South Korea | Canada, Mexico, Asia |
| 761100 | Aluminum tanks or vats for anaerobic digesters for biomass gasification | Not available | Not available | Not available |
| 840681 | Steam turbines (greater than 40 MW) for combined heat and powers and renewable energy applications | New York, North Carolina | Negligible exports | Japan, Germany, China, Belgium |
| 841011 | Small hydraulic turbines and water wheels, and parts | California, Ohio, Oregon, Washington | United Kingdom, Chile | EU, China |
| 841090 | Small hydraulic turbines and water wheels, and parts | California, Ohio, Oregon, Washington | United Kingdom, Chile | EU, China |
| 841181 | Gas turbines, not for use in aircraft, of a power less than 5,000 kW (designed to be used with nonfossil fuels) | California, New Jersey | Neglible exports | Canada, Japan, Italy |
| 841182 | Gas turbines, not for use in aircraft, of a power greater than 5,000 kW (designed to be used with nonfossil fuels) | South Carolina, Texas | Neglible exports | Canada, Italy, United Kingdom |

| HS 6-digit code | Product description | States where there is production and/or assembly | Principal U.S. export markets | Principal foreign producers and exporters |
|---|---|---|---|---|
| Renewable Energy Plant (REP) | | | | |
| 841581 | Geothermal heat pumps | Arkansas, Florida, Indiana, New York, Oklahoma, Texas | Canada | Japan, Germany, Denmark, Germany, China, Mexico |
| 841861 | Geothermal heat pumps | Arkansas, Florida, Indiana, New York, Oklahoma, Texas | Canada | Japan, Germany, Denmark, Germany, China, Mexico |
| 841869 | Geothermal heat pumps | Arkansas, Florida, Indiana, New York, Oklahoma, Texas | Canada | Japan, Germany, Denmark, Germany, China, Mexico |
| 841919 | Solar water heaters and parts | Arkansas, California, Connecticut | Mexico, Antigua Barbuda | China, Germany, Israel, Canada |
| 841990 | Solar water heaters and pars | Arkansas, California, Connecticut | Mexico, Antigua Barbuda | China, Germany, Israel, Canada |
| 848340 | Wind turbine gearboxes | Illinois, Pennsylvania | Neglible exports | Belgium, China, Germany, India |
| 848360 | Flexible couplings for wind turbines | Florida, Minnesota | Neglible exports | Germany |
| 850161 | AC generators for renewable energy applications (with an output not exceeding 75 kVA) | Arkansas, New York, Minnesota, North Carolina, Tennessee, Wisconsin | Not available | EU, China |
| 850162 | AC generators for renewable energy applications (with an output exceeding 75 kVA but not exceeding 375 kVA) | Arkansas, New York, Minnesota, North Carolina, Tennessee, Wisconsin | Not available | EU, China |
| 850163 | AC generators for renewable energy applications (with an output exceeding 375 kVA but not exceeding 750 kVA) | Arkansas, New York, Minnesota, North Carolina, Tennessee, Wisconsin | Not available | EU, China |
| 850164 | AC generators for renewable energy applications (with an output excedding 750 kVA) | Arkansas, New York, Minnesota, North Carolina, Tennessee, Wisconsin | Not available | EU, China |

:

# Annex C. (Continued)

| HS 6-digit code | Product description | States where there is production and/or assembly | Principal U.S. export markets | Principal foreign producers and exporters |
|---|---|---|---|---|
| **Renewable Energy Plant (REP)** | | | | |
| 850231 | Wind-powered generating sets | Arkansas, California, Colorado, Idaho, Iowa, Florida, Pennsylvania, South Carolina, South Dakota, North Dakota, | Canada, Chile, Mexico | Denmark, Germany, Spain, Japan, India |
| 850720 | Deep discharge solar batteries | California, Georgia, Pennsylvania | South America, Africa, India | China, Canada |
| 853710 | Photovoltaic system controllers | California, Colorado, Massachusetts, Oregon | Negligible exports | Germany |
| 854140 | Photovoltaic (solar) system cells and modules | California, Michigan, New Mexico, Ohio, Oregon | EU, China, India, Hong Kong | China, EU, Japan, Taiwan, Malaysia |
| 900190 | Fresnel mirrors | California, New Mexico | Not available | China, EU, Japan, South Korea |
| 900290 | Fresnel reflector modules | California, New Mexico | Negligible exports | China, EU, Japan, South Korea |
| **Environmental Monitoring, Analysis, and Assessment Equipment (EMAA)** | | | | |
| 903210 | Thermostats | Minnesota | Not available | Germany, Mexico, China |
| 903220 | Manostats | Not available | Mexico, Canada | EU, Japan, Mexico |

| HS 6-digit code | Product description | States where there is production and/or assembly | Principal U.S. export markets | Principal foreign producers and exporters |
|---|---|---|---|---|
| Management of Solic and Hazardous Waste (SHW) | | | | |
| 392010 | Polyethylene geomembrane liners and related products | Texas, South Carolina | Not available | Canada, China, Israel |
| 761290 | Aluminum tanks for liquird or solid waste | Virginia, Wisconsin | China, Australia, Canada, Caribbean | China, Taiwan, Japan, EU |
| 840219 | Biomass boilers | Georgia, Nebraska, Texas, Virginia | Not available | Germany, Mexico, China, Canada, Italy |
| 840290 | Parts for biomass boilers | Georgia, Nebraska, Texas, Virginia | Not available | Germany, Mexico, China, Canada, Italy |
| 840410 | Auxiliary plants for biomass boilers | Georgia, Nebraska, Texas, Virginia | Not available | Germany, Mexico, China, Canada, Italy |
| 841940 | Distilling or rectifying plants | Illinois, Massachusetts, Tennessee, Texas, Nevada, New York | Iraq, Spain | Germany |

## Annex C. (Continued)

| HS 6-digit code | Product description | States where there is production and/or assembly | Principal U.S. export markets | Principal foreign producers and exporters |
|---|---|---|---|---|
| Air Pollution Control (APC) | | | | |
| 840490 | Parts of auxiliary plants for biomass boilers | Georgia, Nebraska, Texas, Virginia | Not available | Germany, Mexico, China, Canada, Italy |
| 840510 | Producer gas or water gas generators, with our without their purifiers | California, Colorado, Idaho, New York, Oklahoma | Not available | Germany, Netherlands, China, Japan, Denmark, Canada, Switzerland, Sweden |
| 841989 | Evaporators and dryers for wastewater treatement; condensers and colling towers; biogras reactors; digestion tanks and biogas | Kansas, Maryland, New Hampshire, New Jersey, Pennsylvania, Washington, Winconsin | Canada, Mexico, Brazil, China | EU |
| Heat and Energy Management (HEM) | | | | |
| 701931 | Figerglass mats (insulating) | Alabama, Arizona, California, Colorado, Georgia, Indiana, Kansas, New York, Ohio, Pennsylvania, South Carolina, | Not available | Canada, China, Finland |
| 841950 | Heat exchange units | Colorado, Connecticut, New York, North Carolina, Pennsylvania, Texas | Canada, China, Mexico | India, Germany, China, Denmark |

| HS 6-digit code | Product description | States where there is production and/or assembly | Principal U.S. export markets | Principal foreign producers and exporters |
|---|---|---|---|---|
| Waste Water Management and Potable Water Treatment (WWM) | | | | |
| 560314 | Nonwoven materials for wastewater filtration | Delware, California, Pennsylvania, South Carolina, Texas | Not available | Israel, China, Mexico, Canada, United Kingdom, Japan |
| 730900 | Steel tanks for liquid or solid waste | California, Kansas, Oklahoma | Not available | China, European Union, Japan, Taiwan |
| 732490 | Water saving showerheads | Indiana | Not available | China, Taiwan, Mexico, Germany, Italy |
| Cleaner or More Resource Efficient Technologies and Products (CRET) | | | | |
| 732111 | Solar cookers and ovens | California, Illinois, Maryland, Minnesota | Not available | India, Germany, Mexico, China |
| 732190 | Parts of solar cookers and ovens | California, Illinois, Maryland, Minnesota | Not available | India, Germany, Mexico, China |
| 850680 | Fuel cells | Connecticut, Oregon, New York | Not available | Canada, EU, Japan |

Note.--Product descriptions refer to those environmental goods contained within the broader 6-digit classification description.

## End Notes

[1] Paragraph 31 (iii) (Trade and Environment) states: "With a view to enhancing the mutual supportiveness of trade and environment, we agree to negotiations, without prejudging their outcome, on [...] the reduction or, as appropriate, elimination of tariff and non-tariff barriers to environmental goods and services." WTO, *Ministerial Declaration* (WT/MIN(01)/DEC/1), November 14, 2001.

[2] WTO, Committee on Trade and Environment Special Session, *Proposal for a Result under Paragraph 31 (iii) of the Doha Ministerial Declaration* (JOB(07)/193), December 3, 2007.

[3] Assistance from the U.S. International Trade Commission was led by David Ingersoll.

# INDEX

## D



## E

## F



## G



## H



## I

## P



## R



## S



## T